The Processes of Life

The Processes of Life

An Introduction to Molecular Biology

Lawrence E. Hunter

The MIT Press
Cambridge, Massachusetts
London, England

MIT Press books may be purchased at special quantity discounts for business or sales promotional use. For information, please email special_sales@mitpress.mit.edu or write to Special Sales Department, The MIT Press, 55 Hayward Street, Cambridge, MA 02142.

This book was set in Syntax and Times Roman by SNP Best-set Typesetter Ltd., Hong Kong. Printed and bound in the United States of America.

Library of Congress Cataloging-in-Publication Data

Hunter, Lawrence.
The processes of life : an introduction to molecular biology / Lawrence Hunter.
    p. ; cm.
Includes bibliographical references and index.
ISBN 978-0-262-01305-5 (hard cover : alk. paper) 1. Molecular biology. 2. Computational biology. 3. Evolution. I. Title.

[DNLM: 1. Molecular Biology. 2. Computational Biology. 3. Evolution.
QH 506 H945b' 2009]
QH506.H86 2009
572.8–dc22

2008030905

10 9 8 7 6 5 4 3 2 1

To Jill, Max, and Hailey—my partners in evolution

Contents

Preface

I have to begin with an admission: The only biology course I ever took was in 10th grade, and I was too busy learning the facts of life in the hallways to have paid much attention in that classroom. So I feel obliged to tell the story of how I reached the point where I could write this book, and to thank all the people who helped me along the way.

In college, I was fascinated with the idea that computers could be a useful tool in understanding minds, and struck up what is now a 25-year-old friendship with *l'enfant terrible* of artificial intelligence, Roger Schank. A year after I had graduated, Roger called me back from an idyllic life in rural Hawaii because he had a project that he insisted I come to graduate school to work on. Roger can be very persuasive, and so I returned to lovely New Haven, CT, to meet Ray Yesner, MD, a great lung tumor pathologist, and Jerome Silbert, MD, a colleague of his with a passion for computers. While Roger knew that there was a good PhD thesis in modeling how Ray could organize and remember all that information, he couldn't have suspected that this would lead me to a lifelong love of biology and medicine.

As with some other loves, it didn't start off that well. Important as lung tumor pathology is, I found my thesis work depressing. In those days, despite dozens of distinctions pathologists could make by looking at tumors under a microscope, there were only two therapies to offer patients, and neither of them worked very well. Lung cancer is a horrible disease that kills just about everyone diagnosed with it in short order. It wasn't easy to look at those images, let alone to learn more about what was going on beneath the surface.

Fortunately for me, in addition to introducing me to Drs. Yesner and Silbert, Roger also introduced me to the U.S. National Library of Medicine (NLM), which funded my thesis work. I've always loved libraries, and with freshly minted PhD in hand, NLM was happy to have me as a staff scientist. DNA sequencing technology was advancing rapidly then, and the race to sequence the Human Genome was getting started. For a computer scientist, gene sequences and protein structures and the biological questions about them were much easier to understand (and easier to write useful programs for) than patients and tumors. I worked hard to bring computer scientists and molecular biologists together and help launch contemporary computational biology. In my years at NLM, I met heroic molecular biologists like John Wootton and Harold Morowitz, who seemed to know at least a little something about every gene I ever encountered. Not only were they tremendously helpful colleagues, but their enthusiasm and even some of their knowledge was contagious. I was hooked.

The NLM is on the Bethesda campus of the National Institutes of Health (NIH). The breadth and quality of biomedical research there is probably unsurpassed anywhere in the world. My colleagues at NLM were a pleasure to work with and to learn from. NIH also attracted visits from many great scientists who gave fascinating lectures and showed a surprising willingness to talk with a curious computer scientist. I absorbed all I could. Although it was a struggle at first, once I became comfortable with the language of biology, the rest started to fall into place. As I asked questions of my colleagues, I started to hear the phrase that echoes in my ears to this day, "well, you're about right, but it's a little more complicated than that. . . ." I had a wonderful decade, surrounded by brilliant people and hearing firsthand about many important breakthroughs in biomedicine.

One of the reasons I ultimately left NIH for Colorado was to start a training program in computational biology. When I arrived in 2000, the fallout from the dot-com bust meant there were a lot of well-trained computer scientists looking for something new to do. However, they often struggled in my early computational biology classes, since many didn't know the first thing about biology. At that point, Harvey Greenberg, a friend and mathematician with an interest in things biological, encouraged me to create a course that would introduce molecular biology to the computer scientist. Since he was putting so much effort into creating the training program I wanted to see, I couldn't say no.

Creating that course was an incredibly taxing—and incredibly rewarding—experience. I learned the truth of the aphorism about how you don't really know a topic until you've taught it. All that first semester, I worked late nights and long weekends, often finishing my preparations just moments before class. I had to go back to solidify my understanding of the basics, which I had largely intuited by listening to research presentations, and realized that there was still a great deal of interesting biology I didn't know. Answering students' questions in class was an exercise in humility; my most common answer was "let me look that up and get back to you next class." My teaching assistants that year, Christiaan van Woudenberg and Raphael Bar-Or, gave me far more of an assist than they could have expected. The efforts and enthusiasm of those students kept me going. And my wife and young children made the first of many sacrifices that would ultimately be required of them for me to complete this book.

As I was putting the course together, it became clear that there was no textbook appropriate for it. The course was aimed at mature students who wanted to get up to speed on the whole of molecular biology in a hurry. The typical undergraduate route required years of prerequisites in order to ultimately digest one of the detailed 1,000+ page classic molecular biology textbooks, like *Molecular Biology of the Cell*. That wasn't going to work for these folks. The more gentle introductions, like Hoagland and Dodson's *The Way Life Works*, and Clark and Russell's *Molecular Biology Made Simple and Fun*, were really good, but didn't cover nearly the range of topics I thought the class needed to grasp.

After teaching the course for a couple of years, I came to the conclusion that I really needed to write that textbook. More and more people were interested in adding molecular biology to their professional lives. NIH was encouraging not only computer scientists, but physicists, mathematicians, engineers, ethicists, and others to join in the exciting world of postgenomic biomedicine. Although there was (and remains) a lot of interest, learning enough biology to even get started has been a significant barrier to entry. My goal for this book is to knock down that barrier and make molecular biology accessible to anyone who is seriously curious.

Of course, deciding to write a book is not the same thing as actually writing it, and a lot more people helped me get from the decision to this text. Debbie Kornblith painstakingly transcribed videotapes of my classroom presentations. Boris Tabakoff, my department chair, helped me arrange the sabbatical that it took for me to actually devote enough time to the project to get it done. My entire lab, and particularly my text mining group leader Kevin Cohen, suffered through my absence, and I appreciate all of their efforts; I hope to make it up to you all one day. Helen and Francis Weir and their kids, Gabriel and Dominic, hosted us on their lovely farm in Catalonia and took care of us like long-lost friends. Mary Queally, who owns the cottage in Fanore, Ireland, where I finished the book, was also a wonderful host.

Bob Prior, my editor at the MIT Press, was the right mixture of encouraging and forgiving, and ultimately was very generous with both time and resources. Even with the sabbatical I was two years late on delivering the manuscript, which Bob tells me is not a record. One of the things I am most grateful for is the MIT Press's agreement to allow me to freely distribute the text of the book (without typesetting or figures), as part of a scientific project in text mining. Foresightful and generous publishers are rare, and I am very grateful to be involved with one of the best.

Once the text started flowing, I depended on the advice of many readers to help me improve the book and catch mistakes. My first victims were good friends I have had since those high school days, Alexis Pearce and Thair Peterson. Since neither have any science background (Alexis is a rabbi and Thair is a journalist), slogging through my early drafts was clearly an act of love. Their extensive comments really helped make the book much more understandable. Members of my lab who made substantial comments include Mike Bada, Greg Caporaso, Kevin Cohen, Cheryl Hornbaker, and Philip Ogren. Several faculty colleagues at the University of Colorado Denver School of Medicine, including Marilyn Coors, Michael Holers, Jennifer Richer, and Mark Yarborough, read sections and provided valuable comments. Of course, all remaining errors and infelicities are my responsibility.

Two people made even more substantial contributions to this work. Tzu Phang created wonderful illustrations, working closely with me to capture the spirit of the text in visual form. Mike Bada went through the Glossary using his comprehensive knowledge of biomedical ontologies to link concepts in the book to these increasingly valuable community resources.

Ultimately, the most important help I had in producing this book came from my wife, Jill, and my children, Max and Hailey. They were always supportive, even when it meant I wasn't spending time with them. It can be hard for a five-year-old living in a strange country to keep hearing that daddy has to work on his book, but Hailey was both patient when I was working, and always ready to play when I was done. Max, who has just discovered his own love of writing, was equally supportive, loving, and fun. Jill in particular had to cope with her new business taking off just as we were to leave on the sabbatical. Her offer to take the kids home a month early so I would have the time to finish was above and beyond the call of family duty—and without it, the book might never have been done. Thank you all, from the bottom of my heart.

The Processes of Life

1 In the Beginning . . .

1.1 Approaching the Study of Life

Questions about the origin, functioning, and structures of living things have been pursued by nearly all cultures throughout history. The work of the last two generations has been particularly fruitful, producing a remarkably detailed understanding of how living things operate. This new understanding is grounded in physics and, especially, in chemistry. Insights into the molecules of life have clearly demonstrated how fundamentally ordinary materials can be alive in so many extraordinary ways.

Becoming conversant with the intricacies of molecular biology and its extensive technical vocabulary appears a daunting prospect. Introductory textbooks typically run more than a thousand pages, and college courses in the field can require years of prerequisites. As more and more people become seriously interested in molecular biology, the existing introductory materials too often form more of a barrier to entry than an invitation to the study of life. This book is an attempt to open that door for anyone who wishes to enter.

It's not that molecular biology is more difficult than, say, physics or chemistry, but that the study of life at the molecular level involves so many interconnected strands of knowledge that it is hard to find a good place to start. Life is frustratingly holistic. Studying one organ in isolation from the rest of the body, or even one organism in isolation from all the others, doesn't work well. How is it possible to learn about all of them at once?

Learning molecular biology is like climbing a spiral staircase: one goes around and around a set of core topics (reproduction, evolution, development and so on), each time a topic is revisited, one reaches a higher, more complete understanding. The purpose of this book is to take you around that spiral once, so you are ready to appreciate more detailed knowledge about any aspect you care to pursue further, be it how DNA works or how to treat cancer. A spiral is an imperfect metaphor, since there are so many linkages among biological concepts—for example, how DNA works and treating cancer turn out to be related. It is simply impossible to lay out biology linearly, so this book is laced with cross-references, to help you navigate the connections that didn't fit exactly into the path I chose to get you around the first level of the spiral.

What does it take to go around that spiral once? What are the core topics for understanding life? From at least the days of the early Greeks, humanity has searched for a "substance of life," a special material that was the essence of living things. The search for that special substance turned out to be a mistaken conception of what life is. While some materials (like DNA and proteins) are found in nearly all living things, it is not a special kind of stuff that makes something alive. The mere presence of any particular material (including DNA) doesn't make something alive. The materials of life, it turns out, are just fairly ordinary chemicals, in particular combinations. What makes something alive is not what it *is*, but what it *does*. The *substance* of life is less important than the *process* of life.

To get around one loop of the spiral, you will have to start picking up the terminology of biology, which is a bit like learning a foreign language. The things that molecular biologists talk about don't have a lot of equivalents in the everyday world, so they invented words to describe what they discovered. Learning that language is part of what it takes to understand molecular biology. This book is filled with the technical terms you will need to know to understand other biological texts, each introduced with enough context to make sense. All of these terms are defined in the glossary and **boldfaced** at their first occurrence in the text.

Learning a foreign language involves more than just learning its words and grammar; a language embodies many aspects of its speakers' culture. The same holds true for the language of biology. The culture of biology is different from that of physics or engineering. Biologists conceive of experiments and data in their own ways, and think about the phenomena they study "biologically." Though it is difficult to describe any culture briefly in words, perhaps the most central idea in the culture of biology is the interplay of **structure** and **function**. Structure, to a biologist, describes the details of the physical components of a living system and how they relate to each other; it is what a thing is. A structure can be as complex as an organ (like the brain), or as basic as the shape of a molecule. Function, to a biologist, is the role that a structure plays in the processes of life—what a thing does. Much of what biologists do is to identify the structures that make possible a function, or to identify the function that some structure serves.

Another key aspect of biological culture is its obsession with the particular. Many other kinds of science focus on finding very general rules or laws that describe the behavior of a large part of the universe. Through hard experience, biologists have discovered that there are very few universals in biology. Even some of the most widespread phenomena in life (such as the use of DNA to encode information) turn out not to be quite universal; a few organisms always seem to manage to do things differently. For that reason, biologists are wary of generalizations.

Physicists have long derided biology's lack of generalizing theory. The Nobel prize–winning physicist Ernest Rutherford famously dismissed biology as "mere stamp collecting." He meant that biologists' scientific work has largely been to describe the phenomena of life, in exacting detail. For many years, biological science largely entailed painstakingly cataloging the tiny differences between thousands of kinds of creatures, noting exactly what their bodies look like, how they behave, what they eat, how they reproduce, etc. Physics didn't go around cataloging how different kinds of matter move, but instead created predictive theories of various kinds of motion. In a way, Rutherford was right: There isn't a grand theory of biology that can predict how a robin gets its food, or how yeast reproduce, and there probably never will be.

However, perhaps a better metaphor for the work of those biologists might be collecting biographies, not stamps. Imagine that you wanted to understand what it meant to be human. No matter how well you understood one person, that could not possibly be enough to understand humanity. Learning in detail about a lot of different people's lives, say, by reading many biographies, begins to show something of commonalities and the differences among people, and provides a basis for an understanding of what it is to be human. Different people's lives illuminate different aspects of our shared humanity. No simple theory can provide the same richness of detail, the same fidelity to the essential complexity. This also holds true for our understanding living things. At many points throughout this book, I present a bit of the life story of an organism that illustrates a specific aspect of how life works.

Why collect biographies? Because human beings are different. Why describe so many different organisms in such detail? Because living things are even more different. Life encompasses a tremendously broad range of organisms. There are so many organisms of so many different kinds that it is hard to imagine. The vast majority of organisms in the world are completely outside most people's experience, since they are too small to be seen with the naked eye and live only in a very narrow range of places. The plants and animals with which most people are familiar, diverse as they may be, are really just a tiny fraction of the enormous number of living things currently alive and an even smaller fraction of the life that has ever existed on Earth.

The question of what, exactly, is alive is itself a challenge. No precise definition of "life" is accepted by all scientists as correct. Even the **cell**, the structure that (almost) all living things are made from isn't quite universal: Viruses aren't made of cells.[1] Cells are somewhat circularly defined as the simplest entity that can exist as an independent living system, and they are the basic structural component of all organisms. Many organisms are just a single cell. Human beings are made of more than 100,000,000,000 of them.

However, speaking more informally, it isn't hard to describe some of the essential processes something must do in order to be considered alive. A living thing must **reproduce**, that is, to have the ability to create other organisms (although sterile animals, like mules, are still considered alive even if they cannot reproduce). Questions about reproduction in all its forms, from the functioning of cell division to how the genetic combination of two adults can create a child, are a central topic of biology.

A living thing must also have a **metabolism**, that is, to be able to convert external materials into its own components, or into offspring (although viruses don't have their own metabolism; they hijack the metabolism of other living things to make their components and offspring). Metabolism is a remarkable process. Some organisms, called **autotrophs**, need only inorganic substances (usually carbon dioxide and ammonia) and sunlight or another source of energy to make all of their own components. Some plants and many single-celled organisms are autotrophs. All other creatures are **heterotrophs**, which means they have to consume materials created by other organisms (usually by eating them) to get all the inputs they need to live. All animals, including people, are heterotrophs. We will see how organisms are able to take only available inputs and use them to synthesize the many complex substances—in just the right amounts and organized in just the right way—to create or sustain a living thing, using only ordinary chemistry and physics.

We will explore these and other processes, but the most basic one, the one thing that really does link all living things without exception, is **evolution**. As the remarkable nineteenth-century Jesuit priest and paleontologist Pierre Teilhard de Chardin put it, "Evolution is a light which illuminates all facts." The entire study of life, from biochemistry to ecosystems, hinges on evolutionary explanations.

What is "evolution," anyway? It isn't a particular "theory"; there are many competing theories about how evolution works. Evolution is a kind of process, a way that a particular kind of change occurs. We will look at it in detail in

---

1. Viruses do have to infect a cell in order to reproduce.

the next chapter, but one of the most critical ideas shared by all theories of evolution is both remarkable and easy to understand: all living things are related to each other. All living things have a common ancestor; somewhere, way back when, every creature has a great-great-great-great- . . . -great grand-parent in common with every other creature. Human beings are distant cousins to dinosaurs, and to bacteria; sometime, long ago, we have a common ancestor with both. That's one reason that no one organism can be effectively studied in isolation.

To understand any member of the great family of life, it will help to know about some of the relatives. We will shortly take a look at the diversity of life on Earth, and also the history of living things, before diving into any particular organism in detail.

1.2   Billions and Billions of Creatures . . .

To understand biology, even molecular biology, requires an appreciation of the diversity of living things. There is a tremendous range of differences—in what they look like, how they live, how big they are, where they live, how they eat, how they reproduce, how long they live, what they can sense, what they can do, what they can think, and more. There is variation among species, among individuals within a species, among the parts of an individual organism. From the molecules up to the ecosystems, there is tremendous diversity and tremen-dous variability.

Let's consider a few illustrative examples. Most people looking at an aspen grove see a bunch of separate trees, but it is really a single organism, connected together by a single root system, sharing nutrients and reproducing as a unit. Figure 1.1 shows an aspen grove in Utah, nicknamed Pando, that is generally considered to be the largest organism in the world. It covers more than 100 acres and includes more than 47,000 stems (which look like individual trees). Pando is estimated to weigh more than 6,500 tons, and is likely to be at least 80,000 years old. In contrast, consider the much smaller critter in figure 1.2, named radiodurans. It is an example of a large family of organisms called the *Archaea*, a microscopic form of life that wasn't even suspected to exist 30 years ago.[2] Archaea live in all sorts of environments that were previously thought to be impossible for living things to survive in, like in acid so strong it can dissolve steel, or in superheated, high-pressure volcanic vents at the bottom of the ocean. The radiodurans in figure 1.2 gets its name from its

---

2. Before then, the few that were known were thought to be a kind of bacteria. In fact, despite their small size, they are more closely related to people than they are to bacteria.

Figure 1.1
An Aspen grove. While it appears to be a forest of trees, actually all of the stems visible here are part of the same organism.

amazing ability to survive 100,000 times the radiation that would kill a person. It can also survive being totally dried out—when there is enough water to rehydrate it, it comes back to life. These creatures that survive in such weird places are called *extremophiles* and there are many kinds of them.

There are other kinds of extremes in life as well. Consider the mayfly in figure 1.3. The lifespan of a mayfly is about two weeks, but most of that is spent in a juvenile stage, growing up. Their entire adulthood lasts only about five minutes. In that five minutes, they find a partner, mate, lay eggs, and die.

At the other end of the longevity spectrum is the desert tortoise, seen in figure 1.4. They can live to be over 120 years old; there's one in an Australian zoo claimed to be 177 years old. They may live even longer than that; it's hard to tell when the recordkeepers don't live nearly as long as the tortoises do. The age at which a tortoise reaches adulthood depends on the availability of water, and can range anywhere between 12 and 20 years. Not only do they live a long time, modern turtles have been around in much the same form as they are now for more than 200 million years, more than one hundred

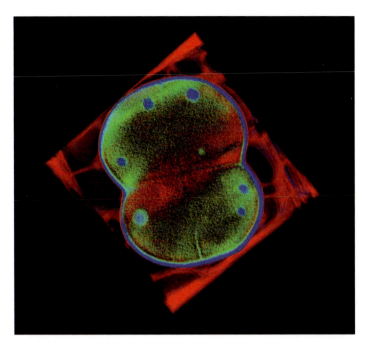

Figure 1.2
Visualization of Cryo-EM tomographic reconstructions of frozen-hydrated *Deinococus radiodurans*. Image courtesy Cristina Siegerist of the U.S. Lawrence Livermore National Laboratory.

times longer than humans. Turtles were there in the very early days of the dinosaurs.

Opabinia, shown in figure 1.5, is a very ancient creature that lived in the Cambrian era, about 500 million years ago, long before dinosaurs. The model in the figure is based on a fossil found in the Burgess shale in the Canadian Rockies, where many beautifully preserved and very ancient creatures were discovered. Opabinia has five eyes and about a third of its body looks like a long flexible tube with teeth at the end. There's nothing at all like it in the modern world, and its fossils demonstrate that there's nothing "universal" about the way life looks now.

These are just a few of the examples of the huge number of species on the planet. How many species are there? It may seem odd, but no one knows exactly. There is some controversy over exactly what should constitute a species. The original definition, attributed to the great biologist Ernst Mayr, is a group of potentially interbreeding organisms that are reproductively isolated from other such groups. However, that makes it difficult to determine species boundaries in organisms that don't reproduce sexually (e.g., bacteria), and the

Figure 1.3
A Mayfly on a white painted board. The fly stands with its wings raised, the black veins in the wings clearly visible. It has three long hairs protruding from the back of its segmented abdomen. This photo was submitted to Free Nature Pictures, http://www.freenaturepictures.com, by Rickard Olsson, with reprint permission under the Creative Commons License.

question of which organisms are "potentially interbreeding" can be hard to answer. We can say with certainty how many species are known[3] and described, and can make some good estimates of lower bounds for other sorts of species. For example, there are 5,416 known species of mammals, 9,934 species of birds, and at least 29,300 species of fish. But those are the familiar and most studied sorts of organisms. Most of the biodiversity in the world is in less familiar organisms. For example, there are at least 258,650 kinds of flowering plants, and 950,000 species of insects. Estimating the number of bacterial **taxa**[4] is extremely difficult. More than 18,000 pure strains are available from the American Type Culture Collection in Maryland, but these are just a

3. One reliable source for species counts, used here, is the International Union for Conservation of Nature and Natural Resources "redlist," http://www.iucnredlist.org/info/tables/table1.

4. Taxa (singular taxon) are groupings of organisms. The idea is similar to species, but more generic. Since exactly what would count as a bacterial species is unclear, I use "taxon" instead. How many there are is still an open topic; see, for example, Thomas Curtis's article "Estimating prokaryotic diversity and its limits," in the *Proceedings of the National Academy of Science*, 99;16 (2002, Aug 6):10494–9, available at http://www.pnas.org/cgi/reprint/142680199v1.

Figure 1.4
A Mohave Desert Tortoise. This is from the U.S. Fish and Wildlife Service, and was taken by Jeff
Servoss.

tiny fraction of what can be found in the world. Recent estimates suggest
there are many millions of kinds of bacteria (see suggested readings at the end
of this chapter), most of which have yet to be studied at all. While no one is
sure, there are probably even more kinds of viruses. Viruses are obligatory
parasites, since they depend on another creature to reproduce. Every organism
also appears to have its own set of viruses, as well. Even bacteria get
viruses.

How to make sense of all this diversity? That question is as old as human
investigation of the living world. Aristotle was among the first to try to organize
living things into categories, that is, to develop a taxonomy of life. In the
mid-1700s, Carl Linneaus devised the first hierarchical taxonomy based on
observable characteristics, and devised a nomenclature (naming system) for
living things that largely persists to this day. The scientific name of an organism has two parts, a **genus** (which names a group of related organisms) and a
species (which identifies a particular member of the group). For example *Homo
sapiens*, the scientific name for people, indicates that we are part of the group
of related organisms called *Homo* (which include *Homo neanderthalensis*, our

Figure 1.5
An imaginary reconstruction of an Opabinia by Olof Helje used by his permission.

extinct relative, popularly called the Neanderthal). Sometimes the genus is abbreviated by its initial letter, as in *H. sapiens*.

There is a much more extensive hierarchy of groupings as well. For example, species in the genus *Homo* are all kinds of primates, which are in turn all kinds of mammals, and so on. Since all of these organisms are related to each other, it is possible to construct a family tree that visually illustrates the groupings and the splits that have occurred. The most basic (and most ancient) split created three very broad divisions: the Bacteria, which are simple, single-celled organisms; the Archaea, which are more complex single-celled organisms containing the extremophiles mentioned previously; and the **Eukaryotes**, which include all plants and animals, and many other organisms as well. Even in the most familiar class of animals, the vertebrates, there are many unfamiliar creatures with a great deal of diversity in aspects such as body plan, environmental niche, and lifestyle. So many organisms are now known that it is impossible to print even a summary of the tree on a single page. An animated tree that is well worth exploring can be found on the wonderful "Understanding Evolution" Web site, http://evolution.berkeley.edu.

1.3 . . . All Alike!

There are more than a million kinds of organisms, each with its own special, defining characteristics: different sizes, shapes, lifespans, ways of eating, ways of reproducing, ways of sensing and acting. How is it possible to make sense of all of this diversity, other than just collecting all the "biographies"? Perhaps the most remarkable discovery in the history of biology is that the molecular processing that goes on in all of these creatures is remarkably alike.

The two most critical functions that living things have to achieve are (1) managing matter and energy so as to stay alive, and (2) creating offspring. The molecular structures that underlie those two basic functions are quite similar in all of these different organisms. This surprisingly uniform set of molecular mechanisms is strong evidence that all living things do, in fact, have a common ancestor, and it is also the reason that so much of biology has become oriented around molecular studies.

The molecules that do all the work of being alive fall into a few, well-defined categories. The first is a set of very large and complex molecules called **proteins**. Proteins, working together in groups, are directly responsible for most of the remarkable things that living things can do, reshaping matter and energy to sustain life and create offspring. Each organism uses thousands of different proteins to go about its life. The ability of radiodurans to survive so much radiation, or of aspen groves to get energy from the sun, each depends on the activity of thousands of particular proteins.

Proteins are interesting molecules, and later we will examine their structure and function in some detail. There are millions of kinds of proteins in the world. The bacterium *Synechococcus elongatus*, a relatively simple autotroph, needs about 2,500 proteins to do its work. Very complex organisms might make 100,000 different proteins, many of which are modest variants of each other. Each protein (a structure) can accomplish a specific function (or sometimes more than one). Proteins can be studied in isolation, but they usually work in groups. A useful metaphor for proteins is as basic commands in a computer programming language; they can be combined in various ways to define all different sorts of processes.

Proteins are very powerful and complex molecules, but they actually have a simple underlying structure: they are **linear polymers**. A polymer is a chemical compound made up of simpler units (called **monomers**) that are repeated many times, like LEGO®[5] blocks. The simple units in proteins are **amino acids**. Most organisms have about 20 different kinds of these pieces. The fact that proteins are linear polymers just means that the parts form no branches or circles: a protein can be thought of as a simple "chain" of amino acid "links." Every protein can be described by enumerating, in order, the amino acids that make up its chain, called the protein's **sequence**. Most proteins contain between 100 and 1,000 amino acids. Protein chains are flexible, and the details of their sequences determine a characteristic three-dimensional shape that each protein folds up into. As described in detail in section 5.1, the particular details of the sequence of amino acids and the resulting folded shape is what gives each protein its unique functional capabilities.

Figure 1.6 shows two representative proteins.[6] Figure 1.6a shows a molecular rendering of part of the protein collagen. Collagen is the most prevalent protein in human beings, making up nearly a quarter of all the protein in your body. It forms strong sheets and cables that support skin, internal organs and tendons, as well as the hard substance that gives shape to the nose and ears. Collagen is a braid of three nearly identical protein chains, each more than 1,400 amino acids long. It is one of the largest proteins in the body. However, this very long chain is made up mostly of the same three amino acids, repeating over and over again. Each of the three amino acids is given a different color in the image, and on one of the chains, the atoms that make up each amino acid are shown as little points. The other protein in figure 1.6b is insulin, one of the smallest proteins in the body, consisting of only 51 amino acids. Insulin is a special kind of protein called a **hormone**. Hormones function as messages, sent from one part of the body to another through the blood, and insulin is one of the most important of those. Its message is about how much sugar is available in the blood, and it plays an important role in managing energy. Insulin, like most other proteins, forms a compact, ball-like shape, called **globular**; collagen is unusual in its extended rodlike shape.

Another important family of molecules common to all living things are the **nucleic acids**, **DNA** (**deoxyribonucleic acid**) and **RNA** (**ribonucleic acid**). The nucleic acids are also linear polymers, although their components are **nucleotides**, not amino acids. There are only four different nucleotides (in

---

5. LEGO® is a trademark of the LEGO Group of companies.

6. Note that the images in both figures 1.6 and 1.7, and, for that matter, any image of a molecule, are really just useful fictions created to help us understand. There are many illustrative ways to portray the structure of molecules, but none is an actual image of a molecule. In figure 1.6a, the molecules are rendered schematically, showing only a few atoms. In figure 1.6b, the schematic molecule illustrates only the bonds between the atoms. In figure 1.7, the atoms are represented as little balls stacked on top of each other. These and other renditions of molecules are described in more detail in section 5.1.1.

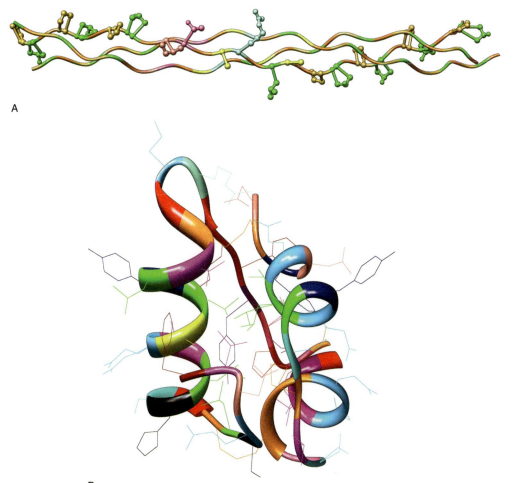

A

B

Figure 1.6
Molecular renderings of two proteins. Panel (A) shows a small portion of the protein collagen, which is a long braid of three nearly identical strands. Panel (B) shows a rendition of a complete insulin protein, one of the smallest proteins found in nature. Section 3.1.2 discusses how to interpret these molecular renditions.

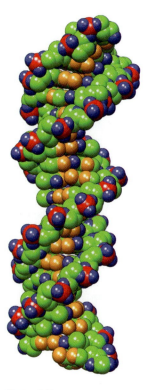

Figure 1.7
A short stretch of a DNA molecule. The green balls represent carbon atoms, the blue represent nitrogen, the red oxygen, and the orange phosphorus. The hydrogen atoms are not shown.

contrast to the 20 or so amino acids). Also unlike proteins, DNA doesn't fold up into different shapes depending on its sequence. Most of the time it forms the well-known "double helix" shape, shown in figure 1.7.

DNA's function is to encode information. More specifically, it encodes all of the sequences of all of the proteins that an organism needs to live. There is a translation between a sequence of nucleotides in a DNA molecule and the sequence of amino acids in a protein. One interesting thing about DNA is that a single molecule of it encodes all of the information necessary to produce thousands of proteins. For that reason, DNA is an extremely long polymer. Some DNA molecules contain more than two hundred million (200,000,000) nucleotides. If you could take one of the DNA molecules in your body and stretch it out straight, it would be nearly three inches long. That is a very long molecule! [The single molecule would be only about 0.000001 of an inch wide, though.]

DNA, RNA, and proteins together are called **macromolecules**. "Macro" means big, since all of those molecules are quite large compared to most compounds that are not produced by living things. In contrast, biologists call all of the other substances involved in life (e.g., sugars, vitamins, oxygen, pharmaceuticals, etc.) **small molecules**. The universality of the structures and functions of the macromolecules, the fact that every living thing uses nucleic acids to encode information and proteins to do biochemical work, is remarkable, considering the diversity in the world of living things. This commonality is what has made possible the advances in understanding life that the past few generations have witnessed.

Molecular biology began with the insight by James Watson and Francis Crick (and also Rosalind Franklin[7]) that DNA was the carrier of the information needed to make all of the proteins in the body. They formulated what is now called the "central dogma" of molecular biology: Information about how to make proteins is encoded in DNA. DNA can be copied to make more DNA, copying all the information in the original. The information in DNA can be transcribed into RNA, which then directs the production of a protein whose sequence of amino acids is determined by the sequence of nucleotides in the DNA (and RNA). The information flow is always either DNA to DNA or DNA to RNA to protein. No other information flow (say, using the information in a protein to create DNA) is possible.

While it is a revolutionary and important way of looking at living things, even the central dogma has a few exceptions. In biology, there are always a few exceptions. In this case, one exception is a family of viruses that actually keeps its protein codes in RNA, and copies information from RNA into DNA, which then directs the host to produce viral proteins and more viral RNA. Since these viruses run the normal process backwards, they are called "retroviruses." HIV, the virus that causes AIDS, is a retrovirus. The retrovirus' ability to use RNA to insert a sequence into DNA turns out to be a useful mechanism in the laboratory, so the mechanism used to accomplish that is now commonly used by scientists and engineers as well (see section 10.3).

The DNA in an organism (RNA for retroviruses) contains all of the information necessary to make an offspring.[8] For that reason, the complete sequence of all of the DNA in an organism can be said to encode its entire heritage. The technical ability to "read" DNA sequences has advanced greatly, and it is now

---

7. See "Rosalind Franklin and the Double Helix" *Physics Today* (March 2003): 61. Available as http://www.physicstoday.org/vol-56/iss-3/p42.html.

8. The proteins in the egg also play a role in the process, but a relatively modest one. The process is discussed in more detail in section 7.2.

Figure 1.8
Phosphatidylcholine, a lipid molecule. The white balls represent hydrogen, the gray carbon, and the red oxygen.

practical to determine the sequence of all of the DNA in an organism. This process is called **genome sequencing**. As will become clearer as this book unfolds, knowing this sequence is very valuable, but we are still a long way from understanding the meaning of all the information that it encodes.

Finally, we will also explore some biologically important small molecules. Perhaps the most important class are the **lipids**, including that most familiar lipid, fat. Lipids are polymers, but they are not linear; sometimes they form branches, as shown in figure 1.8. Lipids play many important roles, including forming the membranes that define the "skin" of individual cells, signaling important messages from one part of the body to another, and providing for the storage of energy. Other important small molecules include sugars, starches, and a ubiquitous compound called ATP (adenosine triphosphate), which is one of the main molecules involved in the distribution of energy.

1.4   Where To from Here

Now we have the concepts to at least say what molecular biology is: the study of the structure and function of biological molecules, large and small. Though the differences among organisms are important, it is now possible to explore the underlying similarities that unite all of life: its molecular mechanisms. The process of life is an elaborate dance of interacting molecules.

To really understand the functioning of those molecules requires looking beyond just the molecular. We start in chapter 2 by looking at the process of evolution, the process that gave rise to both the unity and the diversity of life. Chapter 3 describes just enough chemistry to appreciate the amazing things that organisms are able to do with matter and energy. Building on that foundation of evolution and chemistry, chapter 4 describes the universal processes found in even the simplest life forms, and introduces some of the molecular structures that underlie those functions. Chapter 5 returns to the biological macromolecules, proteins and nucleic acids, explaining in more detail how they carry out the processes of life. Chapter 6 introduces the more complicated structures and processes of the eukaryotes, the branch of life that includes all plants and animals. This chapter covers some of the more familiar—but not universal—processes of life, such as breathing oxygen or having sex. In chapter 7, we consider the complications in making a multicellular organism, including cellular specialization, development, and the molecules that coordinate the activities of many cells together. Chapter 8 describes the anatomy and physiology of animals, focusing on the cardiovascular, immune, and nervous systems. Chapter 9 describes the fundamentals of human disease and its treatment, focusing on infections, heart disease, and cancer. Chapter 10 explains contemporary biotechnology, describing the instruments that make this science possible, and the genetic engineering that is one of the foremost changes molecular biology brings to the wider world. Chapter 11 concludes the book with a brief discussion of bioethics, and a consideration of the profound questions that our growing understanding of molecular biology is raising for society.

Although I have attempted to at least briefly introduce as many central issues in molecular biology as possible, several important aspects of the field were largely left out. Perhaps the most important omission is the failure to describe the experimental methodologies that generated the knowledge described here. Although chapter 10 does touch on some of the instruments used by biologists, the extraordinary diversity of clever experimental approaches is simply too overwhelming to even survey. Learning how to conceive of and evaluate experimental methodology is one of the subtlest parts of molecular biology; absorbing the material here is a necessary prerequisite for those interested in that work.

The other important omission is the downplaying of ecological approaches to understanding life. Living things all exist in the midst of complex communities of other organisms, called **ecosystems**. While brief discussions of the many consequences of living in an ecosystem are scattered throughout the text, many issues that relate to ecosystems as a whole, such as biodiversity, ecosystem services (such as nutrient cycles), and ecosystem dynamics (how ecosystems respond to perturbations) are not discussed, primarily because the molecular basis of these important topics remains largely unknown.

This is an exciting time to be learning about molecular biology. While this book is dense with new ideas and new terms, no background beyond a secondary school education is assumed. With this volume, dedication alone (and frequent references to the glossary) ought to be enough to get a solid grounding in all of the basic material you need to open the door to the exciting world of molecular biomedicine.

2 Evolution

The process of evolution is the explanation for both the amazing diversity of creatures observed in the world and the equally amazing molecular commonalities shared among us all. How can that be? What is evolution?

Evolution is one sort of process of change. Not all kinds of change are evolutionary: the geological growth and erosion of mountains is an example of change that is not evolutionary. Many kinds of change are evolutionary, and the following account touches on controversies about which particular account of evolution best describes the history of life on Earth. However, all of these evolutionary processes have three defining features, as first described by Darwin.

The first aspect of an evolutionary process is what Darwin called "descent," that is, the creation of offspring that inherit characteristics from their parents. Entities that evolve must be able to **reproduce**, that is, make copies of themselves. The copies do not need to be exact replicates, but the offspring need to share at least some of the characteristics of the parent. The characteristics that do transfer from parent to offspring are said to be **heritable**. A particular heritable characteristic is called a **trait**. Inheritance of traits from ancestors is what explains the similarities observed among organisms.

The second aspect of an evolutionary process is some source of variation. If every offspring were always a perfect copy of its parent, there would be no evolution, just a lot of copies. Differences in characteristics among organisms are called **polymorphisms** (Greek for "multiple forms"). One source of these variations is random error, which we call **mutation**. Somehow the copying didn't work precisely, and the offspring came out a little different from the parent. Other sources of variation aren't random, like sexual reproduction. Sexual reproduction intermixes the characteristics of two parents to create an offspring that has some of each parent's traits, in a new combination. This offspring isn't exactly like either parent, but the differences aren't just random changes. Variation in inheritance is what explains the tremendous diversity in the living world.

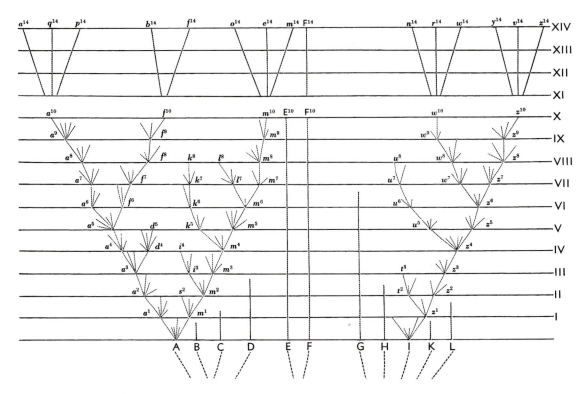

Figure 2.1
Descent with modification, from Darwin's *The Origin of Species*.

The combination of reproduction and variation leads to what Darwin called "descent with modification," one of the central ideas in his book *The Origin of Species*. Figure 2.1 shows the only illustration in that book, which Darwin gave a five-page-long caption to explain how descent with modification can account for similarity and diversity.

Descent with modification doesn't entirely explain the particular characteristics that contemporary creatures have. If variation can be random, why are there such clear distinctions among creatures, rather than a smooth continuum of every possible kind of creature? The third feature of an evolutionary process is what Darwin called **natural selection**, based on the observation that not every creature gets to reproduce equal numbers of offspring. In fact, not every organism gets to reproduce at all. It's entirely likely that the majority of organisms in the history of life didn't have *any* offspring. Some die in infancy, others fail to find mates or have other misfortunes; many difficulties must be surmounted in order to have offspring and become an ancestor. So the organisms that have offspring pass along their heritable characteristics, ultimately determining what particular characteristics were selected frequently enough to be apparent now.

A way of looking at selection from the individual organism's point of view is to define its **reproductive success**, or the number of fertile offspring a creature has in its life. Some aspects of reproductive success are random. Sometimes creatures have more or fewer offspring just by luck. However, some aspects of reproductive success are related to the traits of the organisms, which are then passed on to its descendents. Some heritable characteristics can cause an organism to be more likely to reproduce. The **fitness** of a trait is the contribution it makes to reproductive success. Natural selection is the process by which traits that improve reproductive success—have higher fitness—become more common from generation to generation (and those that hinder reproductive success—have lower fitness—become less common).

Natural selection is a very powerful force. As shown in more detail later, in a **population** (a group of closely related organisms) any variation that makes even a modest or occasional contribution to reproductive success will tend to spread throughout the population over time. A trait that spreads throughout an entire population is said to be **fixed**. Traits that interfere with reproductive success even modestly or occasionally will tend to disappear. It is this loss and fixation of traits in populations that led to the particular organisms living in today's world, rather than a continuum of all possible creatures. Natural selection filters out variations that harm reproductive success, and spreads widely those that improve it.

Many kinds of characteristics influence an organism's reproductive success. How does it get enough matter and energy to live and create an offspring? For sexually reproducing organisms, how can it identify a mate who also has good traits and mate successfully? What kinds of misfortunes must it avoid? The answers to these questions depend on the **environment** in which the organism lives. The environment determines which traits will facilitate reproductive success and which will not. The other living organisms in that environment, called the **ecosystem**, are particularly influential. Darwin emphasized in particular the role that other organisms in the environment play in reproductive success, describing often brutal competition, both within a closely related population and between different sorts of creatures. This emphasis is captured in the full title of his book, *On the Origin of Species by Natural Selection, or the preservation of favoured races in the struggle for life*. It was Herbert Spencer, a contemporary of Darwin, who called this "survival of the fittest."

a diverse interbreeding population with averaging inheritance. Over a modest number of generations, the population would come to be an average of all of the founders. A metaphor that makes this clear is the blending of paint colors: Starting off with a diverse set, blend pairs of colors, then blend the pairs with each other, and so on. After a few generations, all the blends are a uniform color, the average of the original colors.

The solution to that conundrum was first stated by Gregor Mendel, about the same time as Darwin wrote *Origin of Species*, although the importance of Mendel's work wasn't recognized more generally until decades after his death. What Mendel observed was that traits do *not* blend; they are, in his term, **particulate**. Each "particle" of a trait comes from either one parent or the other, without blending. This is the evolutionary concept of a **gene:** the smallest unit of inheritance,[2] something inherited directly from either one parent or the other.

Mendel studied reproduction by controlled mating in pea plants, tracing seven inherited characteristics in more than 10,000 plants over 8 years. The traits he examined were such things as the color and shape of the seeds and pods, the location of the flowers on the stem, and so on. Mendel carefully traced the characteristics not only of hybrids between pairs of plants with particular characteristics, but what happened when those offspring were mated with each other. Through studying many generations of hybrids, he discovered what are now called Mendel's Laws of Inheritance.

Mendel noted that each of the traits he was studying could take on one of a few values. For example, the plant's seeds could be either wrinkled or smooth. We now call those values **alleles**. Wrinkled and smooth are two alleles of the gene that determines seed shape in pea plants. One way to think of this is to consider a gene as a trait that can take on one of several possible values (such as seed texture) and the allele to be one of those values (such as wrinkled).

An allele is not just the value of a trait, however. Mendel hypothesized that each sexually reproducing organism has *two* alleles for each gene: one inherited from each parent. But a seed can't be both wrinkled and smooth; the two alleles need to translate somehow into the observed value for the trait, that is, into the phenotype. An allele that influences phenotype is said to be **expressed**. In the case that the alleles inherited from each parent are the same, there's no question of which will be expressed, and we say the gene is **homozygous**. **Zygote** is the Greek word for egg, and *homo-* means "the same." If the two

---

2. Although the term **gene** was coined by Wilhelm Johannsen in 1909, he was describing Mendel's work.

alleles are the same (say, wrinkled), then the organism's phenotype will express that value (here, have wrinkled seeds).

When the alleles are different, the situation is more complicated. This situation is called being **heterozygous** (*hetero-* means "different")**;** the contributions from each parent are not the same. Since inheritance is particulate, the seeds with both wrinkled and smooth alleles will not turn out to be an average of the shapes, but instead will be one or the other. The allele that is expressed in the phenotype of a heterozygous organism is said to be **dominant**. The one that is not expressed is called **recessive**. The recessive allele does not affect the phenotype, the observable characteristics, of the organism. So, if the smooth seed allele were dominant over the wrinkled one, a heterozygote with a smooth and a wrinkled allele would have a phenotype with smooth seeds. Recessive alleles are observable (affect the phenotype) in only homozygous organisms, when both the alleles are the same.

During meiosis (sexual reproduction), one allele from each parent is selected at random to be passed on to the offspring. For homozygous parents, the result is the same no matter which allele is passed. But for heterozygotes, either allele could be passed to an offspring. When two heterozygous organisms have offspring, each offspring will get one allele from each parent, chosen at random. There are three possible outcomes for that offspring: either it will be homozygous for one or the other allele, or it will also be heterozygous (have one of each allele). The heterozygous offspring will express the dominant allele, like its parents, as will the homozygous offspring with the dominant allele. However, a homozygous offspring with the recessive allele will express that allele in its phenotype. In the observable characteristics of the offspring of heterozygotes, Mendel observed a $3:1$ ratio of dominant to recessive phenotypes.

This theory matched Mendel's experiments well. He started with pure-breed plants; that is, they were all homozygotic for each of the studied characteristics. Then he made hybrids (which had to be heterozygotes) from each pair with pure characteristics, noting that the phenotype of the offspring always matched one of the parents (the one with the dominant allele). Then he bred the hybrids together, and noted the $3:1$ ratio for each of the founding characteristics in this second generation of offspring.

Mendel's insight was that every sexually reproducing organism has an important, but unobservable aspect: what alleles it carries. We now call the complete set of alleles of an organism its **genotype**. The genotype is the inherited contribution to the phenotype of an organism, but it is also more than that. The recessive alleles of the heterozygous genes are not observable characteristics of an organism. Those alleles are hidden in the organism, but they play a role in determining the traits of its offspring. The pair of alleles associated with a single characteristic is often called a **genetic locus** (plural **loci**).

If all of the alleles of an organism are known, then figuring out the alleles of its possible offspring is straightforward. A simple diagram called a Punnett square, seen in figure 2.3, illustrates the possibilities. Capital letters are used for dominant alleles, and lower case for recessive ones. Figure 2.3a shows the possibilities for a single gene. The allele contributed by each parent (arbitrarily labeled paternal and maternal) is combined to produce the combinations shown in the cells, each of which has equal probability. Since any cell with at least one capital letter will produce the dominant phenotype, the observed ratio will be $3:1$. A similar diagram could be constructed to show, say, that when a recessive homozygote (aa) is mated with a heterozygote (Aa), equal numbers of recessive homozygotes and heterozygotes will be produced, yielding a $1:1$ ratio of phenotypes.

Figure 2.3b shows the situation with two different genes, say seed shape and seed color, labeled A and B. Counting up the possible products shown in the diagram, of the 16 possibilities, 9 will have the dominant phenotype for both genes, 3 will have the dominant phenotype A but recessive b, 3 will have the recessive a and the dominant B, and one will exhibit the recessive phenotype for both genes. The $9:3:3:1$ ratio of phenotypes of two characteristics is an indication that they are each controlled by a single gene and that there

Maternal

		A	a
Paternal	A	AA	Aa
	a	Aa	aa

A

Maternal

		AB	Ab	aB	ab
Paternal	AB	AABB	AABb	AaBB	AaBb
	Ab	AABb	Aabb	AaBb	Aabb
	aB	AaBB	AaBb	aaBB	aaBb
	ab	AaBb	Aabb	aaBb	aabb

B

Figure 2.3
Punnett squares. Panel (A) shows the possible combinations of an offspring's alleles for one gene, and panel (B) shows the possible combinations for two genes. Capital letters indicate dominant alleles, and lower case indicate recessives.

is no correlation between the alleles for one character and the alleles for the other; they are said to **assort independently**. Sometimes pairs of genes do show a correlation in their alleles. For example, it might be that white seeds tend to be wrinkled more often than yellow seeds. In that case, there is **linkage** between the genes, discussed in more detail in the next section.

Mendelian inheritance describes **monogenic** traits, that is, ones that are controlled by a single gene. Most traits that people care about (such as a familial susceptibility to heart disease) do not appear to be inherited in a Mendelian, particulate, all or nothing manner. It turns out that such traits are the result of the inheritance of many genes, and are therefore called **polygenic**. Any continuously variable trait, such as height or tendency toward heart disease, must be polygenic. Figure 2.4 illustrates how several Mendelian genes working together can create the appearance of a continuously varying trait, in this case, color. In this artificial example, gene A influences the hue, gene B the saturation, and gene C the brightness.

Mendelian inheritance is a biological function. The physical structure that underlies that function is called a **chromosome**. A chromosome is the part of a cell that embodies its genotype. In most sexually reproducing organisms like people, chromosomes come in pairs, one from each parent. Each chromosome in the pair has the same genes arranged in the same order; the different chromosomes can embody different alleles, of course. The genetic locus includes the corresponding places on both chromosomes. Cells that have pairs of chromosomes are called **diploid**. Some cells (like bacteria) have only one copy of each chromosome. Others (mostly plants) can have more than two copies.

AABBCC	AABBCc	AABbCC	AaBBCC	AABbCc	AaBBCc	AaBbCC	AaBbCc
aABBCC	aABBCc	aABbCC	aaBBCC	aABbCc	aaBBCc	aaBbCC	aaBbCc
AAbBCC	AAbBCc	AAbbCC	AabBCC	AAbbCc	AabBCc	AabbCC	AabbCc
AABBcC	AABBcc	AABbcC	AaBBcC	AABbcc	AaBBcc	AaBbcC	AaBbcc
aABBcC	aABBcc	aABbcC	aaBBcC	aABbcc	aaBBcc	aaBbcC	aaBbcc
aAbBCC	aAbBCc	aAbbCC	aabBCC	aAbbCc	aabBCc	aabbCC	aabbCc
AAbBcC	AAbBcc	AAbbcC	AabBcC	AAbbcc	AabBcc	AabbcC	Aabbcc
aAbBcC	aAbBcc	aAbbcC	aabBcC	aAbbcc	aabBcc	aabbcC	aabbcc

Figure 2.4
An illustration of how several Mendelian genes working together can create the appearance of a continuously varying trait. Genes for hue, saturation, and brightness determine the color of the cells in the figure. Any cell with a capital A is red hued; ones with only lower case (aa) are pink. Any cell with a capital B gets high saturation; only lower case (bb) gets low. Any cell with a capital C gets high brightness of 100; only lower case (cc) gets low.

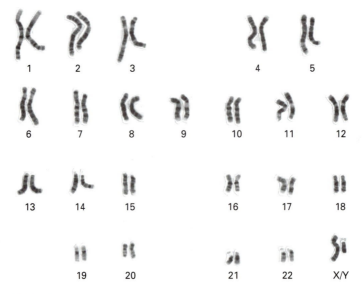

Figure 2.5
A human karyotype, showing 23 pairs of chromosomes. Image courtesy of the US National Human Genome Research Institute.

Some organisms have one pair of chromosomes, but organisms with many genes often divide their genotype into multiple chromosomes. Human beings, for example, have 23 pairs of chromosomes. Mosquitos have 3 pairs; goldfish have 52. Figure 2.5 is a human **karyotype**, which is an image of the chromosomes under a microscope, taken at a particular time in a cell's activity when the chromosomes are separated and easily visible. In the karyotype, the chromosomes are sorted from largest (numbered 1) to smallest (22). The last pair is special, in that it determines the sex of the individual; the one in the image is XY, which is male.

2.2   Variation

In diploid organisms, one chromosome from each pair comes from one parent, and the other chromosome from the other parent. The first step in meiosis is to makes a **germ cell**, also called a **gamete** (sperm or egg). These cells are called **haploid**, since they have only half the usual chromosomes. However, the germ cells don't just take one of the chromosomes inherited from the organism's parents. During the process of creating germ cells, which parental chromosome is the source for the transmitted one alternates every so often, a process called **genetic recombination**. The allele from either parent can get passed along. Exactly when the switch (called **crossover**) happens is arbitrary, but crossovers don't happen too close to each other. The space between crossovers explains linkage, when two genes don't assort independently. Genes that are near each other on a chromosome will tend to be inherited together, getting both alleles from one parent. Genes that are far from each other on a chromosome (or on different chromosomes altogether) will assort independently. Note that when germ cells combine, the new organism gets one chromosome from each parent (which is in turn a mixture of the alleles from that parent's parents), and the fertilized egg is again diploid.

From an evolutionary perspective, this process is remarkably effective at shuffling alleles into different combinations. Not only does the offspring have a combination of alleles from each of its parents, but, because of the recombination during the production of gametes, it also has a combination of alleles from each grandparent (in fact, from many, many previous generations). This helps explain why offspring are different from their parents and other ancestors. Since there are very many possible combinations of alleles, novel combinations (that is, novel genotypes) are likely to arise frequently. Since genes interact in complex ways to create phenotypes, novel phenotypes would be observed frequently as well.

Shuffling of alleles into novel combinations explains how offspring differ from their parents, but sexual reproduction does not provide any explanation of where the particular alleles themselves might come from. One phenomenon that happens rarely, but with significance for evolutionary variation, is a **gene duplication event**. Certain kinds of errors in reproduction of chromosomes can cause a sort of stuttering that inserts an extra copy of a region of the chromosome, sometimes containing one or more genes, into the offspring. As is described in more detail in section 5.2.3, it is possible for an extra copy of one or more genes to play an important role in the evolution of an organism. However, mutation is the only source of genuinely novel alleles, no matter what form of reproduction an organism uses.

The combination of mutation and the allelic reshuffling of sexual recombination seems to many people insufficient to produce all of the remarkable beings in the living world. Many have proposed alternative theories, but none has held up to close scrutiny. Perhaps the most seductive of these goes by the name of Lamarckian inheritance, after the French biologist Jean-Baptiste Lamarck, who did his work well before Darwin published the *Origin of Species*. Lamarck held that characteristics that were used during the lifetime of an organism would be passed on to offspring, and characteristics that were unused would not be passed on. He used the example of a giraffe stretching to reach leaves high on a tree causing the offspring of that giraffe to have longer necks. The idea that an organism striving toward some end would somehow make its offspring more successful is attractive in many ways, among other things, providing a good reason to try to perfect one's self. It just doesn't appear to be true.[3] A closer look at the molecular mechanisms underlying these functions make it even clearer that mutation and recombination are the only sources of variation.

Mutations are the accidental, random changes in genes that create genuinely new alleles. Most mutations either do nothing (called **neutral**) or actually reduce the fitness of an organism (called **deleterious**). Very few mutations are **advantageous**, producing a change that improves fitness. However, as will be demonstrated clearly in a moment, any mutation that is advantageous spreads through a population in a very small number of generations, and deleterious mutations likewise disappear quickly.

The sort of change that happens through the accumulation of advantageous mutants and their spread through a population is sometimes called **microevolution**, in contrast to **macroevolution**, the sorts of changes that occur at or above the level of a species. Macroevolutionary studies address questions like whether the rate of evolution has been relatively steady, or whether there have been brief, rare periods of more rapid changes in the midst of relatively little change (called the **punctuated equilibrium** hypothesis). However, there is no difference in the underlying mechanism involved. Macroevolutionary change is the result of microevolutionary change over large time spans and many populations.

The idea that macroevolution happens as a result of microevolution, that is, that accumulated mutations, just small changes in genotypes, could lead to the amazing and diverse character of the living world is indeed a radical statement, and remains perhaps one of the most difficult conceptual stumbling blocks in learning biology. To make that case here, there is space for only one brief and illustrative example; the suggested readings at the end of the chapter offer much more detailed and careful approach.

One of the challenges in linking together a series of small changes to create a larger one is that each small change, individually, must be advantageous in order to persist. So it is sometimes hard to see how a complex and highly evolved capability could have gotten started. One interesting example is insect

3. Trofim Lysenko, an early Soviet agronomist, argued for a kind of Larmarckian inheritance (and against the newly developing science of genetics) and rose to a position of great power in Stalin's scientific establishment. His vehement opposition to any genetic theory of variation can reasonably be said to have contributed to crop failure and starvation in the early decades of the Soviet Union.

wings, where recent evidence suggests a somewhat surprising origin. Fossils of the first winged insects appeared about 350 million years ago, and many plausible sounding theories about the origins of their wings had been suggested.

Evolutionarily plausible theories about the origin of a complex ability like flight are challenging to produce. Each step along the evolutionary chain that leads to a complex ability must be advantageous in itself. If some structure is a necessary precursor to such an ability, there must be a selective advantage that accrues to the precursor alone. For the evolution of multicellular organisms, there also must be an evolutionarily plausible change in a developmental program that produces the structure (see section 7.1). With respect to wings, very small wings do not provide enough lift to be useful; a larger preexisting structure must have been adapted. What was the precursor, and how did flight arise from it?

Recently, molecular evidence has clarified the issue. Two structures that arose from a common ancestor are said to be **homologs**. Sometimes homology can be assessed by analyzing the physical appearance (the morphology) of a structure: fingers and toes are clearly homologous structures. However, morphology can be ambiguous, and scientists had long disagreed about what structures were homologous to insect wings. Molecular studies of homology, using proteins and DNA to assess common ancestry (see sections 5.1.2 and 5.2.3), are much less subject to alternative interpretations. There is now clear molecular evidence that modern insect wings and a sort of flat plate found in the gills of aquatic insects derived from the same ancestral structure. So, since aquatic creatures with gills had been around for a long time by then, the implication is that wings evolved from some gill-like structure.

Molecular homology resolves only part of the story. All of the intermediate steps between gills and wings had to be independently advantageous. What could those intermediates be? The next clue comes from contemporary stoneflies, a sort of insect that skims across the surface of ponds like a windsurfer. Since their weight is supported by water, skimmers can use the wind to move quickly and in a controlled way, even with small wings and weak flight muscles.[4] As the image in figure 2.6 shows, among contemporary organisms, there is a continuum from swimmers to skimmers to fliers. At this point, a diverse and persuasive body of evidence—some molecular, some functional,

---

4. This is a brief characterization of Jim Marden's groundbreaking work on the evolutionary origins of insect flight. For more information, see his chapter, "Evolution and physiology of flight in aquatic insects," in J. Lancaster (ed.), *Aquatic insects: Challenges to populations* (CABI Press, 2007) or consult his Web pages at http://www.bio.psu.edu/People/Faculty/Marden.

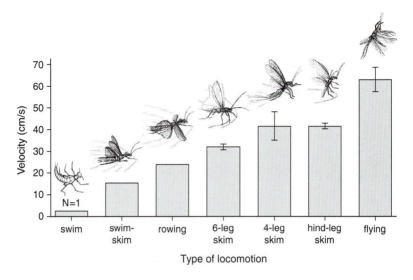

Figure 2.6
A graph showing how the speed of the skimming insect stonefly increases as the amount of contact with water decreases. N in the image represents the number of species that use each form of locomotion, and the pictured insect is representative of one of them. This graphic was created by Jim Marsden, and is used with his permission.

and some structural (e.g. the patterns of veins in related insects)—all supports the idea that insect flight evolved from skimming. Complex abilities can arise through a series of advantageous intermediates, and molecular evidence is tremendously helpful in addressing such questions.

It is also important to note that small changes in a genotype can cause large changes in an organism's phenotype. As chapters 4, 6, and 7 cover in more detail, there are genes whose function is to control the activity of other genes. These kinds of genes play a critical role during the development of an animal from a fertilized egg, and small changes in these genes can change the developmental program, leading to large effects in the adult organism. Figure 2.7 shows two small fish called sticklebacks; an arrow points to the "stickle" coming from the pelvis of the upper fish, missing in the lower one. This significant morphological change is due to a single mutation in the genotype of these fish[5] that effects the development of the body (see section 7.2). These two fish live in different niches, and the other small variations (e.g. the change

5. This story comes from the work of Katie Peichel and David Kingsley in identifying the genetic and molecular mechanisms that underlie evolutionary processes. For more detail, see their chapter, "The molecular genetics of evolutionary change in sticklebacks," in Ostlund-Nilsson, Mayer, and Huntingford (eds.), *Biology of the three-spined stickleback* (CRC Press, 2007), or visit their Web sites, http://www.fhcrc.org/science/labs/peichel/ and http://kingsley.stanford.edu/.

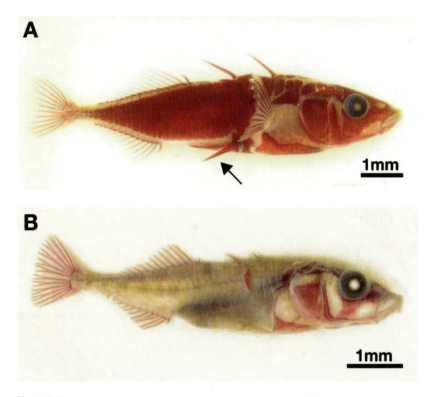

Figure 2.7
Panel (A) shows an Alizarin red-stained stickleback from a Japanese Pacific population. It has large pelvic spines, indicated with an arrow. Panel (B) shows a fish of the same species, but taken from the freshwater Paxton Lake. This fish does not have a pelvic spine, and the dorsal (top) spines are greatly reduced in size. These differences between these two fish are related to a single mutation. This image was taken by Jun Kitano in Katie Peichel's lab, and is used by permission of Wiley.

in color) between them are likely to result from accumulating mutations that are advantageous in their respective environments.

Another way in which a mutation can have a large effect on a phenotype is that genes can play multiple roles in an organism, a phenomenon called **pleiotropy**. Most genes, particularly in animals, are pleiotropic; they play multiple roles, influencing several characteristics. For example, a gene called melanin plays an important role in both giving the color to a cat's fur and transmitting chemical signals from the ear to the brain. Mutations in this one gene can produce cats that are both white and deaf.

Finally, Darwin emphasized the idea of **reproductive isolation** in the creation of new species. The finches he studied were on the Galapagos Islands, and getting from one island to another was difficult for them. The geographic barrier created a reproductive boundary, and the differences among the niches on the various islands create different selective pressures, leading to speciation. This mechanism has now been observed in many specific cases, and is well established. Since Darwin's time, several additional mechanisms of speciation have also been observed. For example, genetic drift (discussed later), resulting from very small populations becoming reproductively isolated from a larger population, can also result in speciation. Another interesting genetic mechanism for the creation of species is diversifying selection, which through the increased fitness of extreme characters (and/or a reduced fitness of heterozygotes) drives a population to split. Examples can be found in tropical insects that become dependent on different host plants, but are not geographically isolated from each other.

2.3   Selection

The third leg of the evolutionary triad is natural selection, the process that determines which genotypes will be successful. In order for a gene to be influenced by natural selection, there has to be a difference in the fitness of the organisms that have various alleles of that gene. Natural selection can function on a gene only if some combination of alleles is better (or worse) than some other combination with respect to reproductive success. Differential reproductive success among individuals with varying genotypes leads to changes in the frequency of alleles (and therefore phenotypes) in a population.

What does natural selection do to allele frequency? Note that this is no longer a question about an individual organism, but about the distribution of alleles (or genotypes) within a population. For a biologist, a population is a group of interbreeding organisms. **Population genetics** studies how the distribution of genotypes changes through successive generations in such a population. The first step in population genetics is to determine what happens when selection is *not* active, when a set of alleles are all equally good in terms of reproductive success. For many years, it was thought that rare alleles would eventually die out, even without selection. However, around 1900 two mathematicians (G. H. Hardy and Wilhelm Weinberg) independently demonstrated that, without selection (and with some other straightforward conditions) allele frequencies in a population are in equilibrium, not changing from generation to generation. Any deviation from that equilibrium would be restored in a single generation.

A population in **Hardy-Weinberg equilibrium** also makes it possible to convert straightforwardly between genotype and allele frequencies. In the simplest case of one locus with two alleles, dominant *A* and recessive *a*, let the frequency of the dominant allele be *p* and the frequency of the recessive allele be *q*. Since there are only two alleles, the sum of these two frequencies must be 1 ($p + q = 1$). Now, think about the genotypes. The frequency of *AA* is just the probability of two independent draws of an *A*, or p^2. Similarly, the frequency of *aa* is q^2, and the frequency of the heterozygote, which can be either *Aa* or *aA*, is $2pq$. Since that covers all the possible genotypes, the sum of those frequencies must also be 1 ($p^2 + 2pq + q^2 = 1$). For a Hardy-Weinberg population, knowing the proportion of homozygous recessives is all that is necessary to calculate all the allele and genotype frequencies. For example, if 1% of such a population has a recessive pheno-type, $q^2 = 0.01$, so $q = 0.10$, which is the frequency of the recessive allele. We know $p + q = 1$, so the frequency of the dominant allele is 0.90; we can then calculate the frequency of the homozygous dominant genotype (0.81) and heterozygotes (0.18) in the population. It is similarly easy to show that these frequencies don't change from generation to generation. These results follow fairly easily from the four Hardy-Weinberg conditions: no selection, no migration in or out of a population, random mating, and infinite population size.

Hardy-Weinberg equilibrium is a useful abstraction, but the assumptions do not apply to any real situation. One of the primary uses of their ideas is comparing observed genotype frequencies to Hardy-Weinberg equilibrium frequencies to characterize the extent to which the assumptions are violated, particularly, to quantify the selective pressure acting on a population. Differ-ences in allele frequency in a population from generation to generation must always be due a violation of one or more of these conditions.

Consider the violation of the population size assumption: Small populations experience a phenomenon known as **genetic drift**, which causes allele fre-quencies to move up or down from generation to generation based on random chance. Genetic drift is a consequence of a sampling bias that is highly depen-dent on the size of the population. In a very small population, alleles can become universal (or completely disappear) entirely by chance.

Migration, or the transfer of alleles from one population to another, is called **gene flow**, and also results in a change of allele frequencies, either adding or removing alleles from a population. The Hardy-Weinberg framework has been used to analyze gene flow in human populations and make interesting histori-cal inferences about prehistoric human migrations (e.g., see *Before the Dawn* in the Suggested Readings).

If organisms select mates on the basis of the degree of similarity (or differ-ence) of their genotypes, it is called **assortative mating**, which violates another

Hardy-Weinberg assumption. For example, if organisms were to preferentially mate with family members (who have similar genotypes) they would become inbred. Inbreeding drives allele frequencies to either one or zero, making such a population genetically uniform.

The violation of the Hardy-Weinberg conditions that is most important in explaining the observable living world is, of course, selection. Unlike violations of the other assumptions, the deviation from equilibrium here is a reflection of fitness, the influence of a genotype on reproductive success. For concreteness, imagine a population of 100 organisms with genotype A, 100 with genotype B, and 100 with genotype C; in the next generation, there are 110 with genotype A, 120 with genotype B, and 90 with genotype C. The **absolute fitness** of a genotype in a population is the ratio of the number of individuals with that genotype before and after selection. In our example, genotype A has an absolute fitness of 1.1, B has 1.2, and C has 0.9. A ratio of greater than 1 means that the genotype is being selected for, and a ratio below 1 means it is being selected against. Using the method for translating between genotype and allele frequencies described above makes it straightforward to quantify the fitness of a particular allele.

Selection can act on the genotypes in a population in several ways. Consider what happens when the homozygous recessive is deleterious, but all the other genotypes are equally fit; this is the case in most Mendelian human diseases. For example, cystic fibrosis is an inherited lung disease, caused by a defective allele of a gene called CFTR. People who are homozygous for the normal allele are fine. People who are heterozygotes, with one normal and one defective allele, are also fine. Since the heterozygotes have the normal phenotype, the normal allele is dominant. However, people who are homozygous for the recessive allele get the disease. It's an awful disease, and most of the people who have it die young, without reproducing.

Since the homozygote recessives don't reproduce, there will be almost none of them in the population. However, since the heterozygotes are fine, they reproduce almost as successfully as the homozygous normals. The only difference is when two heterozygotes reproduce with each other, the Mendelian inheritance means that one out of the four possible offspring genotypes will be homozygous recessive, and that offspring will not reproduce. Even this very small difference in reproductive success of an allele is enough to change its frequency in the population.

The time that it takes for an allele undergoing selection to spread throughout a population depends on the size of the population and the selective advantage of the allele. The formula is

$$T = \frac{2}{f-1} \ln(N)$$

where T is the time in generations, f is the absolute fitness, and $\ln(N)$ is the natural logarithm of the population size, assuming all the Hardy-Weinberg assumptions other than selection hold. So, for a 1% difference in fitness ($f = 1.01$) and a population size of 10,000, it will take a bit less that 2,000 generations for an allele with a 1% fitness advantage to spread throughout the population; it would take the same number of generations for an allele with a 0.99 fitness to be wiped out. So, even if the chances of having fewer offspring due to the bad *CFTR* allele are 1 in 1,000, the proportion of the population with the allele will drop to zero in less than 20,000 generations. The mutation that introduced the bad *CFTR* allele into today's population must have a fairly recent (evolutionarily speaking) origin.

When an allele confers either an advantage or a disadvantage, it is said to be undergoing **directional** or **purifying** selection, which tends to increase or decrease the proportion of an allele in a population. There is also **balancing** or **diversifying** selection, which works to maintain multiple alleles in a population. One kind of balancing selection happens when the heterozygotic state confers an advantage over either homozygote. Sickle-cell anemia, another human genetic disease, has this selective characteristic. Homozygotes for the recessive gene get the disease, which severely affects the life (and reduces the reproductive fitness) of the people who have it. However, the gene is pleiotropic (has more than one effect), and the homozygous dominant has a reduced resistance to malaria, also a devastating disease carried by mosquitoes. So, at least for people living where malaria has a significant effect on fitness, both homozygotes are at a disadvantage compared to the heterozygotes, an example of balancing selection.

Additional sorts of selection can act on polygenic traits. So, for example, directional selection can act to push a trait toward a particular value and away from extremes. Birthweight of human infants is often cited as an example, as there are selective pressures away from weights that are either too low or too high. Similarly, diversifying selection on a polygenic trait can reduce the fitness of a genotype that becomes too frequent (perhaps via parasites or predators), tending to increase the range over which the trait varies.

Fitness can be thought of as a mathematical function that maps from genotype (and environment) to reproductive success. In 1932, Sewall Wright described the idea of a **fitness landscape**, which is a way of envisioning the

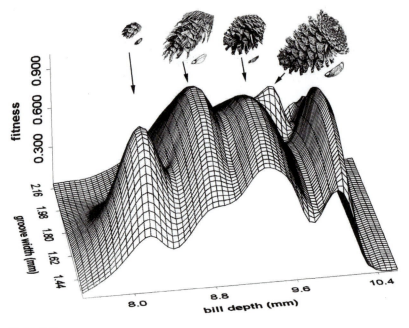

Figure 2.8
A real fitness landscape. The graphic shows the fitness consequences of different aspects of a
bird's mouth (bill depth and groove width): The peaks in the fitness landscape correspond to
mouths best suited to different available food sources. This figure was created by Craig Benkman,
and is used by his permission.

relationship. Each genotype was like a point on a map; genotypes that are
similar to each other are close together on the map, genotypes that are quite
different are far away on the map. The height of the map at a point reflects
the fitness of that genotype. High points in the landscape indicate very successful genotypes, and low points unsuccessful genotypes. An example of the
use of a fitness landscape can be seen in figure 2.8, which illustrates Craig
Benkman's work on crossbills (a kind of bird) showing the relationship
between two aspects of the bird's mouth (bill depth and groove width), the
availability of particular food sources, and the reproductive fitness of the
birds.[6]

The idea of a fitness landscape has been very powerful in understanding
evolution. Think of an organism as a point on the landscape. Variations
in an offspring's genotype are more likely to land on nearby locations in

6. For more details, see Craig Benkman's article Divergent selection drives the adaptive radiation
of crossbills, *Evolution*, 57 (2003), 1176–1181.

the landscape than those far away. Organisms occupying relatively high positions on the landscape will be more reproductively successful than those at lower points. An evolving population (a bunch of points on the landscape) will tend to, metaphorically, climb up the hills. Consider the fitness landscape for the evolution of flight, as discussed earlier. There was a narrow path from one peak (swimming) to another (skimming), to another (flying) without ever moving down (no intermediate forms with reduced fitness).

One clear message of the jagged landscape is that fitness is a very nonlinear function. The fitness of a particular allele depends a lot on the alleles at other loci. If you take just a slice of the landscape, the picture is misleading. Figure 2.8, though not about a single trait, is only slightly less idealized: It shows only two traits, ignoring all the others. Another idealization is the idea of a fixed landscape. The relationship between genotype and fitness changes when the environment changes, so in reality the landscape for a population will shift, perhaps radically, over time. Despite being highly idealized, these diagrams have been very useful heuristics for thinking about evolution.

Other organisms play a central role in defining fitness. Predators, prey, parasites, and competitors greatly influence reproductive success. An adaptation that, say, improves resistance to parasites, or causes problems for competitors, can turn out to be highly advantageous. In most circumstances, the single most important aspect of the environment with respect to fitness is the other organisms that are present. The reciprocity between the fitness of predators and prey, or between hosts and parasites, drives what is called **coevolution**, wherein the evolution of the two species becomes closely linked. **Mutualism** is a relationship between two creatures in which both benefit, such as in the case of clownfish living in sea anemones; each protects the other from predators. The relationship can also be competitive, where one organism benefits at the expense of another.

Competitive coevolutionary relationships, such as those between predators and prey or parasites and their hosts, often lead to cycles of innovation and counterinnovation as the relative success of one leads to increased selective pressure on the other. The interactions between mating partners also play an important role in fitness. As Darwin himself pointed out, **sexual selection**, or the role of mating choices (and of one's attractiveness influencing others' choices) in reproductive success is a particularly important way other organisms influence fitness. Sexual selection accounts for some of the most remarkable features of the natural world, such as the peacock's tail and the bowerbird's nest, shown in figure 2.9.

A

Figure 2.9
Panel (A) shows a peacock's tail, and panel (B) shows a bowerbird's nest, two products of sexual selection. The photograph of the peacock is used by permission of the Oregonian newspaper. The photograph of the bowerbird and his nest was taken by Christopher Zilm, Principal, Toowoomba State High School, Queensland Australia, and is used by his permission.

B

Figure 2.9 (Continued)

2.4   A Brief History of Life on Earth

Evolution is both a mechanism and a set of particular historical events. The variations, fitness landscapes, and selection events that occurred in the history of the Earth all contribute to the characteristics of the particular living things observed today. Coming to know that particular history is a difficult scientific challenge. The main tool for learning about it is the **fossil**.

A fossil is any preserved remnant of an organism. Bones of ancient creatures that have been mineralized (turned to rock) are perhaps the most familiar, but certainly not the only sort of fossil. Fossilized teeth and claws are more common than bones, and can be very informative about the creatures that left them behind. Fossilized embryos and nests are much less common than bones, but provide unique information about the lifestyles of the organisms unavailable from other fossils. Skin imprints, and, very rarely, soft tissues like muscles or organs can be found, usually preserved in amber or lava rock. **Ichnofossils** are remnants of the marks an organism made on the world, rather than the remnants of the organism itself (*Ichno-* is Greek for track or footprint). These include footprints, but also burrows, nests, teethmarks, and the like. **Coprolites**, or fossilized dung, can also provide useful information about what an organism ate and how it lived. Not all fossils are body parts or imprints. **Chemical fossils** are remnants of organisms in the form of compounds that can only have been produced by living things, or in particular isotope ratios[7] that are the product of the activities of living things.

In order to understand the significance of fossils in the history of life, it is important to date them, to figure out when they were made, itself a quite difficult task. Absolute dating methods depend on radioactive decay. A few unstable substances decay into others at a fixed rate, and this can be used to precisely date the age of the sample. For example, about half of the uranium-235 in a sample will turn into lead every 700 million years. By comparing the ratio of lead to U-235 found in a sample, it is possible to estimate how long it has been decaying. However, limits in the ability to precisely quantify really tiny amounts of these substances constrain the method's usefulness in dating a particular sample. For example, so little U-235 decays in, say, one million years, that U-235 dating has very poor resolution for samples that aren't very, very old. On the other hand, dating with the isotope carbon-14 has the opposite problem. Half of C-14 decays into C-13 in 5,000 years. There is so little of it left after about 100,000 years that it is not useful in dating things that are older than that. Since most of the history of life took place in times for which neither C-14 nor U-235 dating is possible, we need other methods to date fossils.

Fortunately, it is also possible to date fossils relatively to each other. Digging down into the Earth shows that rocks were formed in layers, pretty much the same way all over the Earth. It is often possible to identify which rock layer a fossil was found in, and that way give a relative date to fossils found in other rock layers. While the layers are exposed differently in various places, and there are forces that can make assigning layers difficult, many tricks can be used to align layers and assign dates. One interesting one is the use of shark teeth. Sharks are very ancient creatures, with fairly close relatives who have been around since long before dinosaurs. Sharks, even then, had a lot of teeth, and shed them frequently; a typical shark will shed tens of thousands of teeth in its life. Shark teeth also mineralize nicely, so they are widely preserved. Furthermore, the teeth from different kinds of sharks have many small variations, making it possible to compare a single fossil shark tooth to all the many other

7. Atoms and isotopes are explained in section 3.1.1.

shark teeth that have been found to make a very good estimate of the age of the layer and any other fossils found in it. Fossils like shark teeth that are widely distributed are called **index fossils** because they are so useful for dating.

Some isotope ratio methods are good for dating objects back a very long time. The oldest rock ever found can be confidently dated to be about 4 billion years old. Other evidence places the age of the planet at about 4.6 billion years. In its early days, Earth was under intense bombardment by meteorites, which ended about 4 billion years ago. If life had existed before about 4 billion years ago, it would surely have been wiped out by those meteor impacts. They were intense enough, for example, to vaporize any ocean that might have existed at the time. The process of evolution from which we descended could not be any older than about 4 billion years.

Remarkably, the oldest fossil is nearly that old. Chemical fossils, in the form of isotope ratios, have pretty convincingly (although not without some controversy) dated the oldest evidence of life to about 3.85 billion years ago. Even skeptics of the chemical fossils agree that the oldest macroscopic fossils, 3.5-billion-year-old stromatolites found off the coast of Australia, are evidence of organisms somewhat like today's cyanobacteria, which leave behind very similar formations. Regardless of the precise date of the oldest fossil, it is clear that, on geological timescales, life began almost as early as it could have.

The history of life is very long and many organisms have come and gone. This process of evolution informs so much of biology that it is important to know at least the brief overview of the major events in the history of living things. The next few paragraphs recount some of those important moments, but words alone do not convey a good sense of this history. Visiting a museum of natural history, to see real fossils and scientific reconstructions of ancient life forms, is probably the best way to rapidly gain an appreciation. For anyone who hasn't been to a natural history museum since childhood, it is definitely a trip worth taking as an adult.[8]

The most important moment in the history of life was certainly its origin, which, as outlined previously, happened sometime between 3.5 and 4 billion years ago.[9] The life forms that existed then are in some ways similar to some

8. I take my graduate students to our local natural history museum, on the first "field trip" many of them have had since high school. While dubious at first, they almost always find the experience much more enlightening than they thought they would. Try it!

9. The origin of life is a fascinating topic, and molecular biology has illuminated some of the many issues involved. A serious treatment is beyond the scope of this book, however. Useful references include: Pier Luisi's book, *The emergence of life: From chemical origins to synthetic biology*, New York: Cambridge University Press, 2006, and Richard Robinson's article, Jump-starting a cellular world: Investigating the origin of life, from soup to networks, *PLoS Biology* 3 (2005), e396.

contemporary bacteria, whose living processes are discussed in detail in chapter 4. Nearly half of the history of life passed before, from our perspective, significantly different sorts of organisms arose. About 2 billion years ago, the first eukaryotic cells came into being; these are the subject of chapter 6. A billion years ago or so, about three quarters of the way through the history of life, the first multicellular organisms came into being. It was about 580 million years ago that the first very simple animals, a bit like today's sea sponges, came into being. It wasn't until about 540 million years ago, seven eighths of the way through the history of life, that complex animals like the ones most people know of today began to appear. The time period up until that moment is called the Precambrian Era. From about 542 to about 251 million years ago is a period called the Paleozoic (which means "old animals" in Latin). At the beginning of the Paleozoic Era a tremendous diversification in the kinds of animals occurred, so quickly and to such a vast degree that it is called the Cambrian Explosion. Nearly all of the modern classes of animals appeared then. The Paleozoic ended with the largest **mass extinction** in the history of life, about 251 million years ago, when more than two thirds of the species extant at the time died out. This timeline is illustrated in figure 2.10.

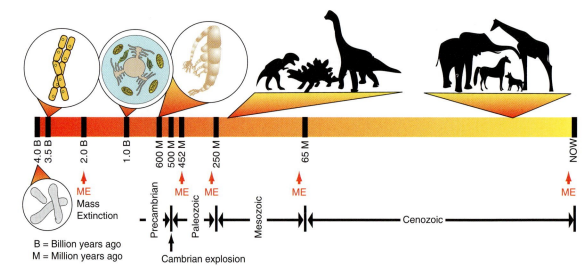

Figure 2.10
A timeline of the history of life on Earth. This is a logarithmic timeline, which means that the more recent events are stretched out and the older events compressed together. The dates of mass extinction events and the boundaries of the eras discussed in the text are noted in the figure, with a few representative organisms shown for each.

Mass extinction events have happened several times. Suddenly, from one rock layer to the next, huge proportions of the species represented in the fossil record disappear. Mass extinctions mark critical moments in evolution. They indicate major changes in the fitness landscape, where organisms that were once well adapted no longer are reproductively successful, and new variations have a better chance of relative advantage. The forces that lead to mass extinction can also reduce the size of populations that are still well adapted. Recall from the earlier discussion that small populations are subject to genetic drift, which can lead to changes in succeeding generations that are not due to differences in fitness.

The number of species on the planet has fluctuated throughout the history of life, and it is possible to make reasonable estimates of it through time. Exactly how many of the life forms of a particular era have to die off to qualify as a "mass" extinction is a bit arbitrary, but a number of events when a large proportion of existing species became extinct in a relatively short period of time clearly deserve the name.[10] The first of these mass extinctions was about 2 billion years ago, caused by what might be called oxygen pollution. In the environment faced by early life forms, there was very little oxygen in the atmosphere, less than 1% (now oxygen forms about 15% of the atmosphere). As will become clearer in section 3.2, oxygen is a very reactive substance. It was produced as a waste product by those early single-celled creatures, and over a very long period of time, the concentration of oxygen in the atmosphere grew. Two things happened as a result: First, creatures that depended on an anaerobic (oxygen-free) environment experienced reduced fitness compared to others that could better tolerate an aerobic (oxygenated) world. Second, variants (described in section 6.2.3) evolved that could take advantage of the reactive oxygen to manage energy much more efficiently, gaining a reproductive advantage over the organisms who had come before. Selection reduced the numbers of the anaerobes and increased the numbers of oxygen users.

The second biggest extinction in the history of life happened much later on, at the end of the Precambrian Era, 452 million years ago. The global temperature dropped quite precipitously, and glaciers covered most of the land. The cause of that event is unknown, but is suspected to be related to volcanism. Very large volcanic eruptions can have devastating consequences for an

10. A timeline of mass extinctions in the fossil record is first documented in D. Raup & J. Sepkoski. Periodic extinction of families and genera, *Science* 231 (1986), 833–836. The "big five" extinctions presented there do not include Precambrian events (since there were no fossils).

entire planet, including reducing sunlight, covering large land masses in lava, poisoning the atmosphere, and changing the weather.

Following that mass extinction, however, was a huge flowering of new species called the Cambrian Era; the first animals that people would recognize as such appeared then. About 200 million years later, that era came to an end as well, this time more certainly due to volcanism. At precisely the same time that the extinction occurred, an enormous volcano developed in Asia. Lava from that volcano covered an area the size of Europe, and the atmosphere was filled with smoke and volcanic gases. Nearly 85% of the ocean species and 70% of the land species were wiped out in a relatively brief period of time.

After that extinction came the Mesozoic (middle-animals) Era, about 250 to 65 million years ago. The Mesozoic saw the origin of all sorts of interesting new classes of organisms, including birds, flowers, and mammals, but is best known for the dinosaurs that lived then. The mass extinction that ended the Mesozoic, and the dinosaurs, appears to have been caused by an enormous meteor that hit the Earth in what is now the Gulf of Mexico, doing immense damage and filling the atmosphere with debris.

The Cenozoic (new animals) is the period from 65 million years ago to the present. The first primates appear about 55 million years ago, and the first monkeys about 30 million years ago. The first member of the human family (hominid), who marks our split from the line that led to chimpanzees, lived about 7 million years ago. The oldest member of our genus (*Homo habilis*) lived about 2 million years ago. The oldest fossil of a modern human is 195,000 years old, and the first evidence of permanent human settlements, agriculture, and symbolic art is about 30,000 years old. Written history began about 6,000 years ago, covering about 0.0000015% of the history of life.

The end of the Cenozoic is marked by the most recent mass extinction event: The one that is going on now. The current extinction is clearly caused by human-induced changes in the environment, particularly the destruction of habitats of many of the world species and geologically rapid changes to the climate. The extent of this extinction cannot be known with certainty for some time to come; reasonable estimates predict that half of the world's species could be gone within 100 years.[11]

---

11. It is beyond the scope of this work to discuss the enormous ramifications of this mass extinction on the Earth and on humanity. For pointers to evidence, news coverage, activism, and other useful resources, http://www.massextinction.net is excellent. Richard Leakey and Roger Lewin's *The sixth extinction: Patterns of life and the future of humankind* is a good book-length treatment of the subject, although now a bit dated.

2.5   Evolution and Life

Evolution is the central phenomena for understanding observations about living things. Biological culture always approaches an observation about the living world by first asking: What role does evolution play?

Evolution is both a process and a particular history. The details of molecular biology are intimately connected to both aspects. Much of what can seem like a complicated morass of details makes coherent sense only in the light of evolution. Knowing a bit about the tree of currently extant organisms, and knowing at least the major landmarks along the timeline of life on Earth makes understanding molecular biology easier.

The most important discovery about life in the last century related the evolutionary idea of the gene to a chemical one. To appreciate it now requires a brief detour into chemistry.

2.6   Suggested Readings

Evolution is a huge topic, with an equally huge literature. The footnotes contain references if you are interested in following up particular topics raised in this chapter. For a book-length treatment of how molecular biology can provide surprisingly detailed evidence about events in human prehistory, see Nicholas Wade's *Before the Dawn: Recovering the Lost History of Our Ancestors* (Penguin Books, 2006). For a deep and philosophical treatment of evolution and its adequacy as a theory, see Daniel Dennett's *Darwin's Dangerous Idea: Evolution and the Meanings of Life* (Simon & Schuster, 1996). Also outstanding and much more accessible, if somewhat dated, is the *Reflections in Natural History* series by Stephen J. Gould. For a fascinating biographical look at one of the great but generally unknown figures in the history of biology, try *Sewall Wright and Evolutionary Biology* by William B. Provine (University of Chicago Press, 1989). There are also many good Web sites about evolution. Two of the best are http://evolution.berkeley.edu and http://tolweb.org.

3   A Little Bit of Chemistry

Evolution is the process of change that shaped living things into the organisms we see today. That process uses no special materials; life is composed of ordinary matter. The characteristics of that ordinary matter have a central role in the evolution of life. The laws of physics and chemistry determine what creatures can exist, determine the possible structures and functions that might be shaped by evolution. Before understanding molecular biology, one must understand something about molecules and the forces that act on them, that is, about chemistry.

Chemistry is an enormous and complex topic, but fortunately only a little bit of chemistry is necessary to get started in molecular biology. The chemistry discussed here shows up at many other points in the book. Before getting to living materials, it is necessary to take a detour into the scientific characterizations of material itself.

3.1   Matter

The substance of life is ordinary matter. A chemist would call most of the substances encountered in everyday life **mixtures**. A mixture is a substance that can be separated into simpler components by physical means (such as grinding or filtering). One special kind of mixture is called **homogeneous**, meaning that any amount of the mixture will have the same composition and properties. A mixture of salt and pepper is most likely *not* homogeneous; there will be places in the mixture where there is a little more salt, or a little more pepper than in other places. Most homogeneous mixtures are liquid **solutions**: the liquid part is called the **solvent** and the stuff dissolved in it is called the **solute**. For example, a spoonful of salt dissolved in a glass of water is a solution; the water is the solvent and the salt is the solute. That solution is homogeneous because every volume of water has the same amount of salt in it as any equal volume of water.[1] Solutions with water as the solvent, called **aqueous** solutions, are particularly important for living things. Almost all of the chemistry relevant to biology happens in aqueous solution.

Chemists have a lot of ways to physically separate the components of a mixture. An important method for biology uses a machine called a **centrifuge**, which takes a vial of a solution and spins it very fast. The force generated by spinning separates the components of the mixture by their density, which is useful for separating solutes in a solution.

Some substances cannot be separated into simpler components by physical means, and so are said to be **pure**. A pure substance is called a **compound**. Although most substances in the everyday world are mixtures, a variety of pure substances are familiar, many of them from the kitchen: water, salt, sugar, and distilled vinegar are all pure. Not all kitchen ingredients are pure (in the chemical sense): flour, milk, and eggs, to name just a few, are mixtures. In fact, the centrifuge was invented to separate milk from cream to make churning butter easier.

Pure compounds cannot be further separated by physical means, but they can be induced to change by a chemical **reaction**. A reaction takes a set of compounds, called **reactants**, and changes them into a different set of compounds, called the **products**. Unlike a physical separation, reactions change the properties of the substances involved, sometimes quite dramatically. The products of a chemical reaction can themselves be separated into pure compounds and then subjected to other reactions. In this way, some compounds can be divided into simpler parts by chemical means. Compounds that cannot be divided into simpler parts by chemical means are called **elements**. There are 94 elements that appear naturally on Earth, and about another 20 that have been synthesized in a laboratory. Of these, only a handful play a significant role in biology.

3.1.1 Atoms

Although it may sound strange, there is a minimum amount of a substance that retains its chemical properties, and cannot be subdivided any further without changing the substance. That smallest amount of an element that retains its properties is called an **atom**. Near the beginning of the twentieth century, it became clear that even atoms had internal constituents. Atoms turned out to be mostly empty space: they contain a tiny **nucleus**, itself made

---

1. For very small volumes and for brief times there may be small differences in concentration, but the solution is still considered homogeneous.

of **protons** and **neutrons**, surrounded at relatively great distance by rapidly moving **electrons**.[2]

These tiny, subatomic particles have only fairly simple characteristics themselves. Each particle has a mass; protons and neutrons have about the same mass, and electrons are much lighter. Protons and electrons (but not neutrons) have an associated electrical charge: protons are positive and electrons are negative. The numbers of protons and electrons in an atom are equal, so the atom itself has no net charge. The particular combination of subatomic particles that makes up an atom determines its properties.

Each element has a unique number of protons, which is a key distinguishing factor. Atoms can gain or lose neutrons, but the atom changes only in its mass, not its other chemical properties. Atoms with a particular number of protons but different numbers of neutrons are called **isotopes**.[3] Atoms can also gain or lose electrons, gaining a net charge. Charged atoms, called **ions**, do have different chemical properties, primarily in the way that they interact with other charged entities, but they also retain many of the properties of the original element; however, an atom cannot change its number of protons without changing its most fundamental properties. If an atom of the inert gas neon were to somehow gain a proton, it would become the highly reactive metal sodium. Because of its central role in determining the characteristics of an element, the number of protons is called its **atomic number**. Each element has a different atomic number, and every atomic number (up to at least 112) is associated with an element.

Atoms also have a mass, which is the sum of the masses of their subatomic particles. The mass of an electron is only about 0.0005 that of a proton or neutron, so the mass of an atom is nearly equal to the number of protons and neutrons. However, each atom often has several naturally occurring isotopes, each of which has a different number of neutrons, and therefore a different mass. The **atomic mass** of an element is the average of the masses of all the isotopes, weighted by their relative abundance. For example, the atomic mass of hydrogen is about 1, and the atomic mass of oxygen is about 16. The name, abbreviation, atomic number, and atomic mass of each element are shown in the **periodic table of the elements**, figure 3.1.

2. An accurate, quantum view of an electron is not like a solid particle moving around at high speed, but instead a probability distribution spread through space. The difference isn't relevant here.

3. Isotopes do have different nuclear properties, such as influencing the time to nuclear decay, but that generally isn't relevant to biology. About the only place where such properties play a role in biology is in the use of isotopic ratios of unstable atoms (those that lose nuclear particles over time) to date fossils; see section 2.4.

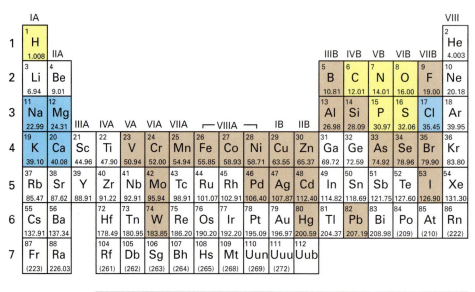

Figure 3.1

The periodic table of the elements. Each box represents an element. Its abbreviation is in the center, the number of protons in the upper left, and the average atomic mass on the bottom. Column labels are related to valence as discussed in the text. The color codes indicate the relationship to living chemistry: yellow indicates the core elements that make up most biological molecules, blue shows commonly exploited ions, and brown indicates elements that are known to play at least a minor role in the chemistry of some organism.

With this background, it is now possible to define "molecule," the key to understanding contemporary biology. A **molecule** is the smallest amount of a **compound** that retains its chemical properties, similarly to the way that an atom is the smallest amount of an element that retains its properties. Equivalently, it is also possible to define molecules in terms of atoms: A molecule is a particular combination of elements bonded together. For example, H_2O is the chemical formula for the compound water, indicating that it is a particular combination of the elements hydrogen (H) and oxygen (O); a water molecule consists of two hydrogen atoms bonded to one oxygen atom. Compounds (molecules) have completely different properties from the elements (atoms) that compose them; water is nothing like either hydrogen or oxygen, despite being composed of those elements.

From the perspective of chemistry, the most important characteristics of an atom (or a molecule) involve its **electronic structure**, the physical distribution and relationships among its electrons. The electronic structure of an atom determines what other atoms it will interact with, that is, what molecular combinations are possible. Atoms bond with each other by sharing or transferring electrons, which results in forces that keep the atoms together, forming a molecule.

Not all of the electrons in an atom are available to be shared or transferred. Electrons are in constant, high-speed motion around the nucleus of an atom following well-defined paths, called **orbitals**.[4] Each orbital has a shape, and they are nested inside each other like layers of an onion. Each orbital can have, at most, a fixed number of electrons occupying it. Once an orbital is full, any additional electrons have to start occupying another orbital. Without going into too much quantum mechanics (you've just done some!), orbitals are grouped into shells, and the number of electrons in the outermost shell, called the **valence**, determines what other atoms a given atom can bond to. The **octet rule** states that each shell can hold eight electrons (except the first, which can hold only two).[5] Each column in the periodic table shown in figure 3.1 is labeled with a Roman numeral between I and VIII, indicating the valence of the elements in that column;[6] for example, the valence of carbon (C) is 4. As the outer shell electrons (that is, valence) of an atom determines how it bonds to other atoms, all of the elements in a particular group have related chemical properties, which is one of the reasons the periodic table turned out to be such a useful tool.

Atoms tend to combine in ways that create full outer shells. Atoms with already full outer shells do not react with other atoms at all; these elements are called inert, for example, the gases helium and neon. Atoms with more than one but fewer than four valence electrons tend to give, or **donate**, their electrons to other atoms. Atoms with valence numbers five through seven tend to take or **accept** electrons from other atoms. Atoms with four valence electrons are equally able to donate or accept electrons, one of the reasons that carbon is particularly important for biology.

The tendency to take electrons from other atoms is called **electronegativity**. Electronegativity is similar to having high valence, although there are other

---

4. From the quantum perspective, an orbital describes the region that has the highest probability density for an electron.

5. This characterization of valence and the octet rule are convenient approximations that work for most biological purposes. Modern valence bond and molecular orbital theory requires more quantum physics than is appropriate here. See the suggested readings.

6. The letter suffixes on the columns are related to the approximate nature of the octet rule.

contributing factors involving the inner orbitals as well, so not all atoms with the same valence have the same electronegativity. Electronegativity is another biologically useful approximation to a more complex underlying quantum reality, and there is no perfect way to calculate it. Linus Pauling developed a widely used scale of electronegativity, ranging from roughly 0.7 (least electronegative) to 4 (most electronegative).

3.1.2   Molecules and Bonds

With the approximations of valence and the octet rule, it is possible to figure out which atoms will bond to which others, and, therefore, which molecules are possible. One kind of bond between atoms occurs when one atom donates its valence electrons to another, resulting in both atoms having complete outer shells. For example, sodium (abbreviated Na) has one valence electron, and chlorine (Cl) has seven. If a sodium atom donates an electron and a chlorine atom accepts one, both would have complete outer shells, so a bond can be formed. The resulting molecule, NaCl or sodium chloride, is better known as table salt. When the atoms give up (or gain) electrons, they become charged. Opposite charges attract, so the atoms involved in an electron transfer are forced together; this is called an **ionic** bond. Ionic bonds tend to occur when there is a large difference in electronegativity between the atoms involved.

When atoms that have similar electronegativities bond, the electrons tend to be shared, rather than transferred outright, creating **covalent** bonds. Covalent bonds work when a pair of electrons is shared between two atoms, as if each got an extra electron in the bargain. Consider two chlorine atoms, each with seven valence electrons and exactly the same (very high) electronegativity. They cannot form an ionic bond. However, when a pair of electrons, one from each atom, is shared, each atom gets a complete outer shell, creating the molecule Cl_2. The sharing of an electron between two atoms makes that electron do a sort of double duty, helping fill the shell of both atoms. A more complicated example is the molecule formed from a carbon atom (valence 4) and two oxygen atoms (valence 6). In this case, two pairs (four electrons) are shared between the carbon and each oxygen. This sharing of two pairs of electrons is called a **double bond**. The carbon atom sees all of the shared electrons, bringing its valence to 8, and each oxygen has one double bond, bringing its total to 8 as well. The covalent bonding between the carbon and two oxygen atoms forms the molecule carbon dioxide, CO_2.

The distribution of electrons within a molecule is also important for its chemical properties. When the electronegativities of two atoms that form a bond are similar enough that the electrons are covalently shared, but are not exactly the same, the electrons will tend to localize toward the more electro-negative atom in the bond. This uneven distribution of electrons results in a **polar** molecule, that is, one where the charge differs in different regions (poles) of the molecule. The larger the difference in electronegativity, the more polar the resulting molecule. Molecules containing atoms with similar elec-tronegativities, and therefore a more even distribution of charge, are called **nonpolar**.

On the Pauling scale, hydrogen has an electronegativity of about 2.2, carbon about 2.5, and oxygen 3.4. So, for example, bonds involving hydrogen and oxygen (2.2 vs. 3.4) will be polar, with most of the negative charge toward the oxygen. In contrast, bonds involving hydrogen and carbon, with their similar electronegativities (2.2 vs. 2.5), will have more uniform charge distri-bution, and are nonpolar.

Because of the attraction between opposite charges, polar compounds tend to be attracted to each other, which tends to exclude nonpolar compounds. The reason oil and water don't mix is that water is polar and oil is nonpolar. Oils are a kind of **hydrocarbon**, a compound made up of carbon and hydrogen atoms. The polarity of water and nonpolarity of hydrocarbons turns out to be very important for living things (e.g., see the discussion of protein folding in section 5.1.1). Nonpolar molecules, like oils, that don't mix well with water are called **hydrophobic**. Nonpolar molecules also tend to aggregate with each other, and exclude polar molecules.

Carbon-containing molecules generally are very important to the chemistry of life; for that reason the study of hydrocarbons was called organic chemistry. Organic chemistry is no longer limited to biological molecules; exotic hydro-carbons such as fullerenes (also called Buckyballs) now fall within its purview as well.

Although there are many more fascinating topics in quantum chemistry, this is enough background to start focusing on the properties of biologically rele-vant molecules and reactions. Each molecule has its own chemical properties. Some of their properties depend simply on their atomic composition: The **molecular weight** of a compound is the sum of the atomic weights of its constituent atoms. For example, the molecular weight of water is a bit over 18, the sum of the atomic weight of oxygen (about 16, see figure 3.1 to look these up) and twice the weight of hydrogen (about 1). However, other chemical properties of molecules are quite different from those of their constituent parts. For example, at room temperature, hydrogen and oxygen are gases, and water is a liquid.

The chemical properties of molecules depend on the precise details of their structures; exactly which atoms are bonded to which other atoms, and also the relationships of the atoms in space. For this reason, it is useful to produce visualizations of molecules in order to understand them. Because they are smaller than the wavelength of light, no one can ever see a molecule, not even with the best possible microscope. All of the visualizations produced are just aids in understanding; none is "correct." There are many useful choices, depending on the context. The simplest characterization of a molecule is its chemical formula, which is a list of the constituent atoms and the number of each in the molecule, such as H_2O. However, these formulas don't convey the spatial relationships between the atoms, nor the bond lengths and the bond angles, which play an important role in the chemical properties of the molecule. For example, some molecules are **chiral**, meaning they have left- and right-handed versions. The chemical formulas are the same, but the molecule can form in either of two mirror image structures. Left- and right-handed versions of a chiral molecule can have different chemical properties.

Figure 3.2 shows five different visualizations of the simplest hydrocarbon, methane, which has the formula CH_4. Figure 3.2a, the "stick" model, clearly shows the tetrahedral layout of the four hydrogen atoms around the carbon; the lengths and angles of the bonds are accurately portrayed. In figure 3.2b, the "space filling" model, which represents that the atoms themselves take up

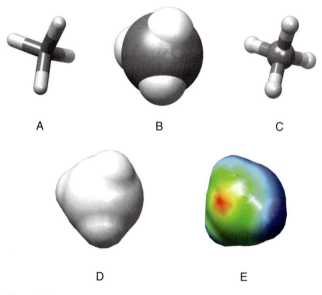

Figure 3.2
Five different renditions of the molecule methane. See the text for a discussion of each rendition.

space, each atom is drawn with a fixed radius. The advantage of the space filling model is that it gives a sense of the molecule's surface, which is most important in determining how it interacts with other molecules. The drawback of a space filling rendition is that the bond lengths and angles are somewhat obscured, and the fixed atomic radius is an imperfect approximation of the surface. Figure 3.2c, the "ball and stick" model, is a hybrid of the previous two, giving a sense of the different size of the atoms, but also showing the bond details. Figure 3.2d shows the calculation of the boundary that 95% of the electron density around the molecule falls within, more accurately capturing the molecular electronic structure. Such renditions are also useful for envisioning the surface of the molecule. Figure 3.2e color-codes that surface to show the distribution of charge (blue for negative, red for positive), which is also important in determining its interaction with other molecules. This is perhaps the most informative rendition, but it also takes the most computation to create. Each of these ways of looking at a molecule can trigger different understandings, and they can all be useful in one circumstance or another.

Another important thing to remember about representations of organic molecules is that they often leave out the hydrogen atoms, showing only the heavy (i.e., nonhydrogen) atoms. These implicit hydrogens are left out of visualizations because there are often so many they would clutter the image, and are easily inferred. If it seems as if a representation of a carbon atom in a molecule is missing some of its bonds, assume there are enough hydrogens attached to ensure every carbon has a total of four bonds (paying attention to any double or triple bonds, of course).

Bonds between atoms create molecules. These bonds are very stable but change during chemical reactions, as described in the next section. However, the same forces that play a role in creating atomic bonds also act on molecules and atoms that are not bonded to each other, causing specific sorts of interactions among molecules. The nonbond effects of these forces on molecules are important in understanding the biological properties of and interactions among molecules.

Charge distribution on the surface of a molecule is important because of the **electrostatic force**, in which like charges repel and opposite charges attract each other. When the charges are near each other and no other forces act in opposition, this is the force that creates ionic bonds. When the force acts on molecules, the distances are greater and the individual atoms are often covalently bonded. Nevertheless, electrostatic force drives molecules (even specific parts of molecules) toward or away from each other. The strength of the electrostatic force between molecules is less than that which holds together an ionic bond, but it still is one of the strongest forces acting on molecules.

Other forces that act on molecules play important roles in biology. One such force is the **Van der Waals** force, abbreviated VdW, which is the only attractive force between neutral (uncharged) atoms. From a quantum perspective, the VdW force is actually the sum of four independent intermolecular forces, but for biology it suffices to view VdW as have two components: an attractive force at long distances, and a repulsive one at short distances. The force acts as if molecules (and atoms) have a preferred distance apart. If they get too close, they push away, but as they get further away than the preferred distance, an attractive force pulls them closer together. This force drops off with distance, so the atoms have to be fairly close together to feel much attractive force. VdW forces are much weaker than electrostatic interactions. Often the size of the atoms in space-filling visualizations is set to be the VdW radius, that preferred distance where the VdW forces are balanced.

A third force acting on molecules, intermediate in strength between electrostatics and VdW, is the **hydrogen bond**. Hydrogen bonds are an attractive force between electronegative atoms (ones that tend to take electrons) bonded to a hydrogen atom (called **donors**), and other electronegative atoms (called hydrogen bond **acceptors**). Hydrogen bonding is as if the two electronegative atoms were sharing the hydrogen. Hydrogen bonds are like a much weaker version of a covalent bond, analogous to the relationship between electrostatic force and ionic bonds. For example, oxygen is highly electronegative, so water molecules form extensive hydrogen bonds among each other. These weak bonds are constantly being made and broken. Carbon has enough electronegativity to be a hydrogen bond donor, and hydrogen bonds play an important role in biology. The relatively low strength of these bonds can add up when there are many such interactions. For example, a large number of hydrogen bonds hold together the two strands of the DNA double helix.

3.2   Reactions

Molecules are subject to chemical changes through reactions. Living things take compounds found in the world and convert them into offspring, so which reactions are possible under what circumstances is a central issue. Chemical reactions can unite two molecules to create a third, split a molecule into two smaller parts, or substitute a part of one molecule for a part of another. Which of these reactions are possible, and under what circumstances, is one of the most important topics in chemistry. It is also central in biology.

One of the simplest sorts of reactions is called **dissociation**, when a molecule splits into smaller parts. When molecules held together by an ionic bond (e.g., table salt, NaCl) dissolve in water, they dissociate. This is because a solution is a homogeneous mixture, so the solute molecules are spread apart and become surrounded by water molecules. The polar water molecules have charged parts, so there is an electrostatic attraction between the water molecules and the atomic charges that are holding the molecule together. This attraction is called the **solvation effect**, and for an ionic compound, that means it will dissociate in water. The transferred electron stays with the acceptor (the chlorine atom, Cl, in the example), so each of the dissociated atoms has a charge, and is therefore an ion. The acceptor gets the electron, and hence a negative charge, and is called an **anion** (pronounced an-eye-on). The donor (sodium, or Na, in the example) loses one of its electrons, becomes positively charged, and is called a cation (cat-eye-on). Solutions of ionically bonded compounds become solutions of charged ions of the atoms that made up the compound. Salt dissolved in water is a mixture of Cl^- and Na^+ ions. This reaction is fully reversible: When the solvent evaporates, the ionic bonds will be reconstituted.

Most chemical reactions involve multiple compounds. Another simple reaction combines two or more compounds to form a third. Burning is the combination of a combustible substance with oxygen, for example, $H_2 + O_2 \rightarrow H_2O$. That reaction is the burning of hydrogen gas (a diatomic—two-atom—molecule) with oxygen gas (another diatomic molecule) to form water.[7] One of the fundamental rules of chemistry is the conservation of matter: The amount of matter in the products has to be the same as in the reactants. When something is burned, whatever mass is lost by the burned substance must be conserved by creating an equal amount of exhaust. For that reason, there is a problem with the preceding example: Two atoms of oxygen go into the reaction, but only one comes out. Coefficients must be added to the compounds in the equation so that matter is conserved; this is called balancing the equation. In this case, the correct reaction is $2H_2 + O_2 \rightarrow 2H_2O$, which means that twice as many hydrogen gas molecules are needed as oxygen gas molecules, and there will be as many water molecules produced as there were hydrogen molecules. The quantitative ratio of the compounds in a reaction is called its **stoichiometry** (stoy-kee-om-et-tree).

Translating the molecular ratios in a reaction into physical amounts of the compounds involved is a little tricky. A gram of hydrogen doesn't have the same number of molecules as a gram of oxygen. To address this problem, chemists invented a useful measurement for the amount of a compound, called

---

7. Burning hydrogen results in water, one of the reasons a hydrogen-powered car is attractive. The reason water itself doesn't burn is that it is the product of burning hydrogen; it's already burned.

a **mole**. A mole of one substance has the same number of molecules as a mole of any other substance. Two moles of hydrogen combine with one mole of oxygen to produce two moles of water. How much does one mole of a substance weigh? A mole of a substance is the amount that has the mass in grams equal to the compound's molecular weight; for example, one mole of water (H_2O) is about 18 grams, a mole of hydrogen gas (H_2) about 2 grams, and a mole of oxygen gas (O_2) about 32 grams. Check the atomic weights of each in figure 3.1 to see why. So, if we burn 2 moles of hydrogen in the presence of 1 mole of oxygen, we will get 2 moles of water. Note that the amount of matter is conserved in grams as well: 4 grams of hydrogen burned with 32 grams of oxygen produces 36 grams of water. A mole is a fixed number of molecules for every substance, but a very large one. There are about 6.022×10^{23} (602,200,000,000,000,000,000,000) molecules in a mole. This constant is known as Avogadro's number. Molecules are indeed very small!

Chemical reactions generally involve multiple reactants and multiple products, but the same rules apply. Consider the burning of methane: $CH_4 + 2O_2 \rightarrow CO_2 + 2H_2O$. Note the stoichiometric constants necessary to balance the equation. If 16 grams of methane (CH_4) were burned, how much oxygen would be needed, and how much carbon dioxide (CO_2) would be produced, in grams? The periodic table provides the answers.

Unlike burning, many reactions are **reversible**, particularly ones that take place in aqueous solution. Starting off with some amount of reactants and some amount of products, the reaction goes in both directions (sometimes forward, where reactants turn into products; sometimes reverse, where products turn into reactants) until a steady state, called **equilibrium**, is reached. Reversible reactions are written with a two-headed arrow, like this: $aA + bB \rightleftharpoons cC + dD$, where the lowercase letters are stoichiometric constants and the uppercase letters are molecules. Every reaction has an **equilibrium constant**, called K_{eq} so that

$$K_{eq} = \frac{[C]^c [D]^d}{[A]^a [B]^b}$$

where each term in square brackets [X] is the concentration of molecule X in the solution.[8] Each reaction has its own K_{eq}, which can change with changes in temperature or other factors. When K_{eq} is large, there is much more of the

---

8. This treatment is a simplification of equilibrium constants that ignores gas phase, activity versus concentration, etc., but is fine for most biological purposes.

products at equilibrium than of the reactants, like an irreversible reaction; when K_{eq} is about 1, there are about equal quantities of reactants and products at equilibrium. When K_{eq} is very small, the reaction doesn't happen very much at all; equilibrium is mostly reactants, with very little product. In calculating this equilibrium, concentrations are figured in moles per liter, called the **molarity** of a solution, with unit M. If 29 grams of table salt (NaCl, molecular weight about 58 as calculated from figure 3.1) were dissolved in a half-liter of water, the result would be a one molar salt solution (1/2 mole of salt divided by 1/2 liter of water), written 1 M.

Each reaction also has a **reaction rate**, which can vary widely. The oxidation of iron (rusting) can take years. The oxidation of butane (burning) takes place in a small fraction of a second. The reaction rate is expressed as a negative number for reactants being consumed, and a positive number for the products produced. The rate is defined for each compound, as the change in concentration over time divided by the compound's stoichiometric constant. Chemical **kinetics** is the study of factors related to the rate of a chemical reaction.

The rate of any particular reaction can also be affected by other factors, such as the temperature and the concentration of the reactants. For biological systems, one of the key factors is the concentration of the reactants, which can be very small. When the concentrations are so low that a reaction happens only very slowly, the reaction is said to be **kinetically unfavorable**.

Even water itself can dissociate, into H^+ (that is, a hydrogen cation) and OH^- (an anion called **hydroxide**), although the equilibrium constant is tiny— 0.0000000000001, or 10^{-14}. With a little algebra, it is not hard to calculate that at equilibrium the concentration of both ions is 10^{-7}, and the concentration of H_2O is almost completely unchanged.[9] The concentration of these hydrogen and hydroxide ions, even though it is small, turns out to be very important. Solutions with more H^+ ions than water at equilibrium are called acids; those with fewer (and therefore more OH^-) are called bases. The interactions of acids and bases are another important part of the chemistry of aqueous solutions, and therefore of life.

The tiny number 10^{-7} is inconvenient to work with for a variety of reasons, so chemists prefer to work with the negative of its log ($-\log_{10} 10^{-7} = 7$), indicated by using a lowercase p prefix. pH, the familiar measure of acidity, is the negative log of the concentration of hydrogen ions in solution. A solution with

9. Yes, water has a concentration: There are 1,000 grams of water in a liter, and the molecular weight of water is about 18, so the concentration of water in solution is about 55.5 M.

100 times as many hydrogen ions as water at equilibrium has an H^+ concentration of 10^{-5} and a pH of 5; that's a mild acid. Since chemists take the *negative* log of the concentration, a *lower* pH means *more* hydrogen ions, and a stronger acid. Compounds that release hydrogen ions when they dissociate in water are called acids; for example, H_2SO_4, sulfuric acid, releases two hydrogen ions per mole, and is a strong acid. By virtue of the extra electron acceptors (hydrogen ions) in acids, they often react strongly with other compounds.

Another important sort of reaction involves the transfer of an electron from one molecule to another. These are called **oxidation-reduction** reactions, or **redox** for short. The compound that donates an electron is **oxidized** (and becomes more positive), and the one that gains an electron is **reduced** (becoming more negative). This change in the oxidation state changes the reactivity of the compound, as if the compound had changed valence. This has a major effect on the chemical properties of a compound; for example, rust is oxidized iron. Oxygen is a powerful oxidation agent; it takes electrons from other compounds, oxidizing them (hence the name). It is this property that made oxygen toxic to the single-celled creatures that died off in the first mass extinction (see section 2.4). However, there are oxidizers other than oxygen, too. Redox reactions are important in photosynthesis and other biological processes that manipulate energy.

3.3   Organic Chemistry

Of the more than 100 known elements, the primary constituents of biological macromolecules include only six: carbon, hydrogen, oxygen, nitrogen, phosphorus and sulfur. A handful of others play roles as ions: iron, chlorine, sodium, zinc, magnesium, potassium, and calcium.[10] The small number of elements found in biological molecules does not make them simple, however. Many biological molecules have thousands of these atoms in them, and small differences in the particular atoms and bonds in such a molecule can have a profound effect on its properties. Organic chemistry focuses on understanding the relationship between these particular molecular structures and their properties, that is, their functions.

A carbon atom, with its four valence electrons, can make bonds with up to four other atoms. If one of those bonds is to another carbon atom, the two carbon atoms can together bond to six other atoms. If one of those bonds is to a third carbon atom, the three carbons can now bond to eight other atoms,

10. A few other metal ions are occasionally used by living things, although in tiny concentrations, e.g., selenium, copper, nickel, molybdenum.

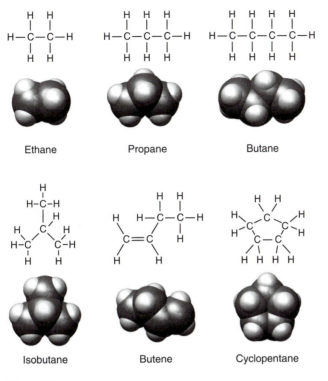

Figure 3.3
Six different hydrocarbons, shown as both chemical structures and space-filling renditions.

and so on. **Hydrocarbons** are the molecules that occur when those other atoms are all hydrogen: CH_4 is methane, C_2H_6 is ethane, C_3H_8 is propane, and C_4H_{10} is butane. As shown in figure 3.3, these molecules look like chains of carbon molecules. Recall that carbon atoms can also form double bonds (share two pairs of electrons) including with another carbon atom, as the example of butene in the figure shows; the double bond is represented by two lines connecting the bonded atoms. Hydrocarbons that have as many hydrogens as possible attached to the carbons (i.e., do not have any double bonds between the carbons) are said to be **saturated**. Propane is a saturated hydrocarbon; butene is not.

An interesting possibility arises when there are four or more carbon atoms in the chain: it is possible for one of the carbon atoms to be attached to three other carbons, rather than each being bonded to two. A carbon bonded to three atoms creates a branch in the chain. So there are two ways to assemble a molecule with the formula C_4H_{10}, one branched and one not. These two molecules have different chemical properties, even though they have the same formula. Molecules with the same chemical formula but different bonding patterns are called **isomers** of each other; the branched version of C_4H_{10} is called isobutane. With at least five carbon atoms, it is possible to attach the ends of the chain together, forming a ring of carbon atoms. Molecules with rings, such as cyclopentane shown in figure 3.3, are called **cyclic**.

Plain hydrocarbons, just carbon and hydrogen atoms, don't do much other than burn. It is the addition of other kinds of atoms that creates the enormous variety of organic compounds, many with biologically useful properties. Biologically important molecules such as proteins and nucleic acids often have chains that involve tens of thousands of atoms, most but not all of them carbon and hydrogen. These are very large molecules, but they obey the same rules as the smaller examples considered so far.

Small groupings of bonded atoms within a larger molecule, called **functional groups**, often imbue the molecules that contain them with a characteristic set of properties. For example, a hydroxyl group (OH), attached to one of the carbons of a hydrocarbon chain makes it an alcohol. Figure 3.4 shows that when a hydroxyl group is attached to one of the carbons in ethane (from figure 3.3), replacing one of its hydrogen atoms, it becomes ethanol, or the sort of alcohol people often drink. A hydroxyl group can also be added to methane (making methanol), which has a related set of properties (e.g., both methanol and ethanol are liquids at room temperature, but ethane and methane are gases). Even though the two alcohols have some similar properties, there are also differences: for example, methanol is acutely toxic to humans. Another important functional group for life is **phosphate**, $-PO_4$. Phosphate is a more complex functional group than hydroxyl: it has one double bond (between the phosphorus and one of the oxygen atoms) and a charge of -3. Bonds among multiple phosphate groups, such as diphosphates (two phosphates) or triphosphates (three phosphates), contain a lot of energy and play an important role in living systems (see section 3.4).

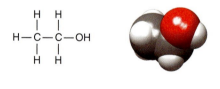

Ethanol

Figure 3.4
The chemical structure and a space-filling rendition of the molecule ethanol. The OH group, shown in red and white in the figure, makes this molecule an alcohol. Ethanol is the sort of alcohol that people drink.

Most of the chemical reactions that make a difference in biology involve the addition, substitution, or removal of a functional group. When talking about functional groups themselves, the place where the rest of the molecule would be attached is sometimes written as just an R, not unlike the role that "x" sometimes plays as a variable in algebra. For example, an organic molecule with a hydroxyl group can be written as ROH. Another set of important functional groups are the ones that make organic molecules into acids or bases. The hydrogen atoms bonded to a carbon do not dissociate in water, so they do not create acids. However, many organic compounds will donate hydrogen cations, making them acids. One particularly common acidic functional group, called the carboxyl group, is a carbon double bonded to an oxygen, and also singly bonded with another oxygen, which in turn is bonded to a hydrogen (−C=OOH). Figuring out which functional groups are present in a molecule can be confusing. Although there is an OH within an RC=OOH, the group functions as a carboxyl, not an alcohol.

Organic molecules can be grouped into broad classes based on their overall structure and the presence of important functional groups. Proteins and nucleic acids were introduced in section 1.2 and will be elaborated on in detail in chapter 5. Two simpler classes or organic molecules, lipids and carbohydrates, are introduced here.

Lipids are distinguished by having a head end that is polar, interacting well with water, and having a long tail that is nonpolar, which tends to be excluded from water. The simplest sorts of lipids are the **fatty acids** that are hydrocarbon chains with a carboxylic acid (C=OOH) group at one end; several of these are shown in figure 3.5. The acid end is polar (and can form hydrogen bonds). Fatty acids can be saturated, meaning they have no double bonds between the carbons and there are as many hydrogen atoms attached to the molecule as possible; **monounsaturated**, meaning they contain a single double bond between a pair of carbon atoms; or **polyunsaturated**, meaning more than one double bond is present. In figure 3.5, the saturated fatty acids appear straight, the monounsaturated ones appear to have a bend or kink, and the polyunsaturated ones have multiple bends or turns.

Three fatty acids can be linked together at the head ends by a three-carbon compound called glycerol, making a **triglyceride**. Most fats are triglycerides, the main molecules animals use for energy storage. Glycerol can also link two fatty acids and a phosphate group into a **phospholipid**, one of which plays an important role in cell membranes.

Carbohydrates, also known as **saccharides**, are molecules with many hydroxyl (OH) groups bonded to carbons and are often cyclic compounds. The simplest are called monosaccharides, and include glucose and fructose (fruit

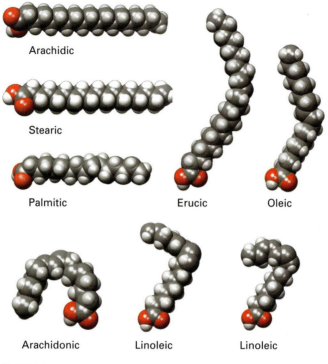

Figure 3.5

Eight different fatty acids. These are hydrocarbon chains with a C=OOH group at one end (the oxygen is rendered in red). The straight chains are said to be saturated; the chains with one bend are monounsaturated, and the ones with multiple bends are polyunsaturated.

sugar). These simple units can be combined in a large number of ways, for example, paired to create disaccharides like sucrose (table sugar) and lactose (milk sugar). Larger groups of saccharide units are called polysaccharides, and include starches, glycogen, and cellulose. These molecules can contain thousands of monosaccharide units, and are used by living things for energy storage and other functions.

3.4   Energy and Thermodynamics

Living things must manipulate not only matter, but also **energy**. Like matter, energy is conserved, and cannot be created or destroyed. However, energy can be transferred from one place to another. Concentrated forms of energy can be harnessed to do work, like moving or heating something. The study of how energy is concentrated, used, and dissipated is called **thermodynamics**.

One of the key ideas of thermodynamics is that of a system. Defining a thermodynamic system is a way of drawing boundaries, of separating insides (the system) from outsides (the system's surroundings). Nearly anything can be viewed as a thermodynamic system. The key requirement is that it be possible to quantify all of the matter and energy that enters or exits the system. A test tube, a living thing, or the entire planet Earth can all be viewed as thermodynamic systems. Systems that exchange matter and energy with their surroundings are called open; all living things are thermodynamically open systems.

The total energy of a system is designated U.[11] U is the sum of the kinetic energy of all of the matter inside the system and all of the potential energy of the interactions among the particles. Oddly, the total energy of a system cannot be measured, so the quantity U can't be determined. However, changes in U, written ΔU[12] can be measured.

There are only a few ways to change the energy of a system: adding or removing mass, heating or cooling it, and having it do work (or doing work on it). The first law of thermodynamics formalizes that idea: The change in the total energy of a system is equal to the amount of energy added by matter entering and by heating, minus the energy lost through matter exiting and in the form of work done by the system. In biochemistry, we can often assume that no matter flows in or out of the system, and just look at the change in total energy as the heat going into the system (designated q) and the work done by the system (designated w): $\Delta U = q - w$. For example, some chemical reactions take work to make them happen, so there has to be a source of heat to make the reaction take place. Reactions that require heating to happen are called **endothermic**. Other sorts of reactions do work on the surroundings; they generate heat, and are called **exothermic**. Burning methane, for example, is highly exothermic.

The heat content of a system is called its **enthalpy**, denoted with H. Like total energy, it cannot be measured directly, but changes in it, called ΔH, can be measured. In a chemical reaction, the difference between the heat of the products and the heat of the reactants is the change in enthalpy for that reaction. Exothermic reactions have positive ΔH, and endothermic reactions have negative ΔH.

Another fact about energy is that concentrated forms of it are more able to do work than dispersed forms, and there is a tendency for energy to disperse

11. This is really the total energy that can be transferred as heat; some forms of energy, such as the nuclear energy holding the atoms together, isn't counted in U.

12. The Greek letter Δ (pronounced delta) is used as a generic indication of change, in order to avoid calculus. Some of its uses are therefore inexact in the following treatment.

over time. The second law of thermodynamics formalizes that idea by defining **entropy**, the amount of energy in a system that *cannot* be used to do work, and to state that any process through which a system does work will also increase the amount of entropy by at least as much as the work done. Formally, $\Delta U = T\Delta S - w$, or the change in system energy is equal to the temperature of the system times the change in entropy minus the amount of work done.

The idea of entropy (energy not available to do work in a system) can be used to define a more biologically useful term, the energy that *is* available to do work: the **Gibbs free energy**, or G. By doing a little bit of algebra, we can see that $\Delta G = \Delta H - T\Delta S$, or the amount of energy available to do work in a system depends on its enthalpy (the heat content of the system), minus the product of temperature and entropy (the amount of energy that cannot be used to do work). The second law also implies that unless matter or heat is added to a system, its free energy has to stay the same or decrease.[13]

Looking at chemical reactions as thermodynamic systems turns out to be quite useful. One use is determining which chemical reactions are **thermodynamically feasible**. No chemical reaction can create more energy available to do work than it started with; there is no perpetual motion machine in chemistry. That is, if a reaction would result in a positive ΔG, it will need an input of at least that much energy to make it happen. The ability to distinguish feasible from infeasible reactions is one of the main uses of thermodynamics in biochemistry.

Just because a reaction is thermodynamically feasible doesn't mean that it will happen spontaneously. Recall the example of the combustion of methane with oxygen used in the previous equilibrium calculation. If methane and oxygen are combined at room temperature, nothing happens. The equilibrium calculation suggests that nearly all of the reactants should be turned into products. The reaction is quite exothermic, and thermodynamically feasible. So why does methane not just spontaneously combust when in the presence of oxygen? The answer is that there is a transition state that the system must pass through, called the **activation barrier**, which has a positive ΔG. Such a transition state acts as a barrier to the reaction; the energy required to get over it is

---

13. This is the basis for the idea that the second law of thermodynamics means that everything becomes more disordered over time. That statement is true only of closed systems (ones with no heat or matter flowing into them). Perhaps the only genuinely closed system is the entire universe, which indeed must become more disordered over time. Everything else, including the entire earth, can become more ordered over time, as long as there is sufficient input, say, from the sun.

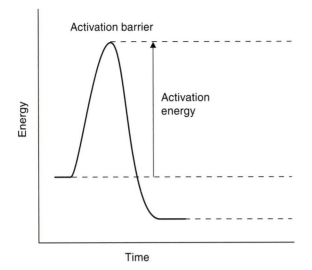

Figure 3.6
A graphical representation of activation energy, the amount of energy required to get a reaction started. In this illustration, the initial energy of the system (shown at the far left) is higher than the final energy (shown at the far right), so the reaction is thermodynamically feasible. However, for the reaction to take place, additional energy is required to get over the activation barrier and initiate the reaction.

called the **activation energy** of the reaction. The combustion of methane with oxygen has a very small activation energy; just a tiny spark is enough to get the reaction started, and it will then go to equilibrium. Figure 3.6 shows the idea of activation energy graphically. Living things often exploit reactions that require activation energy to get started, and, as section 4.2 describes, proteins often provide that critical input.

Another useful feature of thermodynamics is that it doesn't matter *how* a system changes state, only the states themselves matter in the calculations. For example, it is possible to define a molecule as a thermodynamic system and look at the energy involved in making or breaking a single bond within that molecule, without having to consider what reaction made or broke the bond. Every bond among the atoms in a molecule has its own ΔG, called the **bond energy**. The bond energy is the amount of useful work that could be liberated (or is required, depending on the sign) when the bond is broken, regardless of the reaction that breaks it. Living things, which need to break and make bonds to refashion the molecules found in the environment into self and offspring, must manage the energy required to do that work.

Living things are, quite remarkably, also able to make thermodynamically infeasible reactions occur when necessary. There's no magic involved. Instead, the infeasible reaction is coupled to another reaction that produces enough energy to make the system thermodynamically feasible as a whole. Living things have a variety of ways of bringing the needed energy to bear on these reactions, but one of the most common ways is to break one of the phosphate bonds in a molecule called **adenosine triphosphate** (ATP). The energy in that bond is enough to be able to do significant work, but not so much as to inadvertently break bonds in other molecules that the organism needs. ATP also provides the energy that proteins use to overcome activation barriers. One of the most central sets of chemical reactions in all of life is the one that takes either sunlight or food from the environment and produces ATP, effectively "charging the batteries" that power most biological activities.

3.5 Chemistry and Life

The management of matter and energy are central processes of life. Transformation of available matter into offspring requires the use of many chemical reactions. Most of those reactions require input of energy. Energy is needed for overcoming activation barriers for even feasible reactions, and even more energy is required to couple with infeasible reactions to make those possible. Furthermore, any system that reduces entropy (concentrates energy, or, colloquially, increases order) has to get energy externally to balance that reduction in entropy.

The flow of energy through living things is what makes life possible. Each living thing needs to be thermodynamically coupled with a rich source of energy. For many creatures, this source of energy is sunlight. For others, it is electrons from the strongly reducing chemistry of deep-sea volcanic activity. For the rest of us, our source of energy involves eating energy-dense molecules (e.g., fats and sugars) produced by other living things. Limitations on available energy, and competition among organisms for access to it, has driven evolution to make all living things extraordinarily energy efficient.

The chemistry of biological molecules addresses many deep questions about the structure and function of living things. The field of biochemistry has produced countless insights into how living things make the apparently miraculous transformations of matter and energy into offspring within the constraints of the laws of chemistry. This brief treatment of chemistry, in conjunction with the coverage of evolution in the previous chapter, is enough to set the stage for a molecular understanding of simple creatures, to which we now turn.

3.6 Suggested Readings

Both *Chemistry for Dummies* and the *Complete Idiot's Guide to Chemistry* are good starts on chemistry; there's also a two-volume *Organic Chemistry for Dummies*. These are plain English guides, containing a modest amount of mathematics. *Organic Chemistry as a Second Language*, by David Klein, is also popular. A more serious book is *Chemistry: An Introduction to General, Organic, and Biological Chemistry*, by Karen Timberlake. The 9th (current) edition comes with a useful CD-ROM as well.

4   The Structure and Function of Bacteria

Bacteria, also sometimes called **prokaryotes**, are a good place to begin the study of molecular biology, since they have a relatively simple structure compared to other living things. They may be simple, but bacteria are by far the most reproductively successful creatures on Earth, now and throughout history.

Of the total mass of all living things (the planetary **biomass**), more than half is bacteria. These creatures occupy just about every niche that can be imagined; they live in deep-sea oil deposits 20 miles beneath the surface of the earth, and high up in the last fringes of atmosphere bordering outer space.

A human body has about 1,000,000,000,000 cells in it; 90% of those are bacteria, not human cells.[1] Most of these bacteria are either **commensal**, meaning they are neutral with respect to our fitness, or **mutualists**, which provide a benefit. Only a few are **parasites** that harm people.

Though people often think of bacteria as disease-causing germs, and a few are, many are also helpful to people. People couldn't eat many of the foods they do without a flourishing community of intestinal bacteria. Many foodstuffs, like cheeses, are made by letting certain kinds of bacteria grow in the ingredients. However, the vast majority of bacteria don't interact with people at all. In fact, only recent developments in molecular technology have allowed scientists to discover the enormous number of different sorts of bacteria in far more places than had ever been suspected. Most of these newly discovered organisms have barely begun to be characterized.

Biologists study an organism's structure and function at many levels of detail. The first step, at the whole-organism level, is to carefully describe

---

1. Bacterial cells are much smaller than human cells, though. Between 5,000 and 10,000 species of bacteria live in the human body. See Cynthia Sears, A dynamic partnership: Celebrating our gut flora. *Anaerobe*, 11 (2005), 247–251.

observable features of each creature; this work is akin to collecting biographies. The next questions are about its functions, what the organism is able to do: How does it eat, reproduce, and react to changes in its environment? How are the activities of the organism related to its fitness? The structural questions are about what the parts of the organism are, how they relate to each other, and, sometimes, how they develop. Biological explanations link structures to functions, describing the processes and other mechanisms by which evolutionarily advantageous adaptations are realized. Each answer opens new questions about the structures and functions that give rise to the phenomena.

Bacteria are so successful because they can reproduce like crazy. Under ideal conditions, some bacteria can reproduce every 20 minutes. At that exponential growth rate, a bacterial population (called a **colony**) could increase a millionfold in less than 7 hours. Rapid reproduction has high fitness. If two kinds of bacteria were otherwise identical, but one reproduced even just a few minutes more quickly than the other, over a relatively brief period of time, the faster reproducer would come to dominate the population. Unsurprisingly then, many of today's bacteria are very efficient reproducers.

Bacteria reproduce mitotically, without any genetic reshuffling from two parents. Bacterial offspring are, at least in the absence of mutations, genetically identical with their parents. Organisms that have precisely the same genotype are called **clones**, and a population of genetically identical bacterial is called a **clonal colony**. Because there is no sexual recombination in bacteria, the only source of the variation in bacterial genotypes is mutation.[2]

Some bacteria are **autotrophs**, which means they can live on only inorganic materials, like carbon dioxide, ammonia, and sunlight. How these bacteria can live this way provides some insight into how the very earliest life forms might have been able to do the same. Some of these bacteria have quite small genomes as well; the bacterium called *Synechococcus elongatus* demonstrates that as few as 2,500 genes are enough to make a modern autotroph.

The two most central functions that the bacterium must carry out are to reproduce and to maintain itself using the available matter and energy, a process called **metabolism**. All other functions must support these two. For example, some bacteria exhibit **chemotaxis**, meaning that they effectively swim toward food and away from toxins, a function that straightforwardly relates to metabolic needs. Biologists often conduct experiments to test hypoth-

---

2. Infrequently, bacteria do exchange DNA, in the form of free-floating chromosomes called **plasmids**. This is an important mechanism for the spread of drug resistance among disease-causing bacteria. When bacteria (or other organisms) get genes from peers, rather than parents, it is called **horizontal transmission**. This mechanism turns the "tree" of descent into a network of branching and rejoining genotypes.

eses about how particular observed functional capabilities relate to reproductive success, or about how particular structures are able to create those capabilities.

Structurally, a bacterium is a single **cell**. A cell is the simplest structure that supports independent life (some parasites, like viruses, are simpler, but they depend on the cellular structures of other organisms to reproduce). All living things are made of cells, and bacterial cells are relatively small and simple compared with the cells of many other organisms.

Understanding a biological structure requires knowing its components, their relationships to each other, and how the structure achieves the functions it supports. All cells have at least three components: a boundary, which divides the inside of the cell from the outside world; a genome, which contains the organism's genes; and **cytoplasm**, which is the name for the complex mixture of proteins and other compounds that supports metabolism and, in combination with the genome, reproduction.

4.1   Membranes and Boundaries

The **cell membrane** is the boundary that distinguishes the inside of the living thing from the environment outside. Cell membranes are made of **phospholipids**, two fatty acids covalently linked by glycerol with a phosphate group attached to it (therefore called a phosphoglyceride). Cell membranes consist of two layers of these phospholipids, organized so that the fatty acid hydrophobic (excluded from water) ends face each other on the inside of the membrane, and the polar (interacts with water) phosphoglyceride ends face out, one to the cytoplasm and one the environment, as shown in figure 4.1. The molecules shown in green in the figure are water, which dominates both the inside and the outside of the cell, but is largely excluded from the membrane, making that membrane a very effective barrier to flows of matter.

Cells need to get particular sorts of matter in and out, so membranes are **selectively permeable**. Since an organism may need to internalize or emit different substances at different times, controlling the flow of matter through boundaries, like posting a guard at a gate, was evolutionarily advantageous. The structures that support this function are called **pores** in the membrane. Pores are proteins through which only specific ions or molecules can flow. **Channels** are pores with **gates** that can be opened and closed, precisely controlling internal concentrations of particular inputs or wastes. Sometimes energy is required to pump particular substances across the membrane, a process called **active transport**. The membrane is also studded with other proteins, called **receptors**, which react to the presence of nutrients, toxins, and

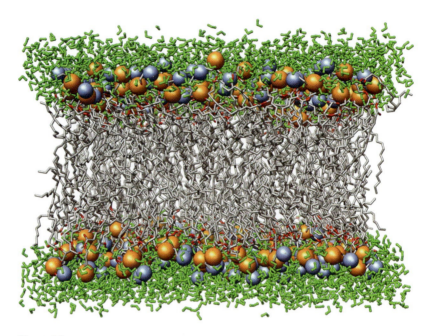

Figure 4.1
A simplified rendition of a cell membrane. The gray shows the hydrophobic tails of the phospho-glycerides, with their phosphates and glycerols shown in blue and gold (red indicates the oxygen atoms at the head of the fatty acids). Water molecules on the inside and outside of the membrane are shown in green; they are mostly excluded from the inside of the membrane.

other external conditions important to the bacterium. Some bacteria also have external appendages anchored in the membrane (**flagella**) that allow them to propel themselves or cell adhesion molecules that help them stick to things in the world.

Since the concentration of ions and other substances is different on the inside of the cell than on the outside, there is a force, called **osmotic pressure**, that pushes against the cell membrane. Bacteria counter this force by building a rigid outer wall around the cell membrane, made out of peptidoglycans, which are saccharides tightly cross-linked with bits of protein. This wall also gives bacteria their characteristic shape (some are rodlike, others spirals, etc.). Without a cell wall, the osmotic pressure would cause the bacteria to burst (called **lysis**, the rupturing of the cell membrane). Many antibiotics work by disrupting bacterial cell walls.

Bacterial walls can be further surrounded by specialized structures that manage the bacterium's interaction with the outside world. For example, some bacteria create an external slime layer that makes it difficult for other organisms to eat them.

4.2 Cytoplasm and Metabolism

The cytoplasm is the structure that supports most of a bacterium's functioning. It contains many molecular systems, specialized for the sort of niche the bacterium lives in. Most of this activity involves the chemical breakdown and synthesis of molecules, termed **metabolism**. Metabolic activities are split into **catabolism**, which breaks down materials available from the environment, and **anabolism**, which uses the resulting building blocks of matter and energy to create the materials needed for self and offspring. Many of these activities, called **core metabolism**, are very similar in all organisms. The metabolic activities seen only in a specialized subset of organisms are called **secondary metabolism**. One core metabolic activity is to tap some source of energy, which could be sunlight, sugars, or other high-energy compounds in the environment, or even redox electrons from compounds emitted by underwater volcanic vents, and transform it into the form needed by the rest of the cell.

The form living things use most frequently is the bond energy of one of the three phosphate groups of a molecule called adenosine triphosphate, better known as ATP. When the bond holding one of those phosphate groups to the rest of the molecule is broken, the reaction releases that energy, as well as a free phosphate and a molecule of adenosine *di*phosphate (ADP). The process that metabolizes food creates the chemistry required to bond a free phosphate to a molecule of ADP, "recharging" it into ATP, transforming food energy into the form the rest of the organism can use.

No single chemical reaction is adequate to convert food energy into ATP, so living things employ a series of linked reactions called a **metabolic pathway**. In a pathway, the products of one chemical reaction become the reactants in the next, linking the reactions together. Through a series of steps, some initial compound is transformed into a different, biologically useful compound.

Metabolism is a large, interacting collection of linked chemical reactions. However, almost none of those reactions happen spontaneously. Each has a required activation energy, many have a very slow reaction rate, some are thermodynamically infeasible, and some have a combination of all three. Nevertheless, living things make these reactions happen, and quickly, through the use of **catalysts**. A catalyst is a substance that increases the speed of a chemical reaction or lowers the required activation energy, without being consumed itself. A catalyst cannot, by itself, make a thermodynamically infeasible reaction take place, although it is possible for one to couple such a reaction with another reaction, usually involving ATP, which provides enough energy to effectively make the otherwise infeasible combination of reactions feasible.

Biological catalysts are called **enzymes**, and the reactants on which they operate are called their **substrates**. Most enzymes are proteins,[3] and they differ from other kinds of catalysts in that they are extremely specific, catalyzing only one very particular reaction on one substrate (or perhaps on a small family of substrates). How the structures of proteins enable them to function as such powerful and specific catalysts is explained in section 5.1.

Metabolic pathways involve a large number of linked reactions, each one of which is catalyzed by an enzyme. These pathways may seem to involve lots of unfamiliar names and subtle chemical transformations, but they are the essence of how organisms manage to convert available matter and energy into offspring and are worth taking the time to understand. If the example of energy metabolism presented next seems overwhelming, review sections 3.2, 3.3, and 3.4, and use figure 4.2 to help navigate through glycolysis.

Energy metabolism begins with a source of high-energy electrons, such as a sugar molecule produced by another organism, an inorganic molecule like hydrogen sulfide, or as the result of photosynthesis (discussed in section 6.2.4). A series of reactions called an **electron transport chain** stores that energy in ATP molecules by capturing the energy in each electron bit by bit, transferring the electron from a donor, through a series of spatially separated redox reactions, to an electron acceptor. Each intermediate reaction uses a portion of the energy of the electron to pump hydrogen ions[4] out through the cell membrane. The hydrogen ions are osmotically driven back across the membrane to regain equilibrium, and that energy is converted into ATP. The enzyme that converts the flow of hydrogen ions into ATP is called **ATP synthase** (enzymes are often named for their function, and generally have an *-ase* suffix to indicate that they are enzymes). Different bacteria have varying electron transport chains, exploiting a wide variety of electron donors and acceptors, dependent on the environmental circumstances of the organism. Despite these differences, the final step of using osmotic flows of hydrogen ions to power ATP synthase is found in nearly all organisms, suggesting that it originated very early in the history of life.

One representative example of an energy metabolism pathway, called **glycolysis**, takes an environmentally available sugar and uses it to charge up two molecules of ADP into ATP (it also charges up two other energy-related molecules, called NAD; more on that in a moment). The glycolysis pathway

---

3. RNA can also act as a catalyst; when it does, it is called a **ribozyme**; see section 5.2.2.

4. Since a hydrogen atom is just one protein and one electron, when the electron is stripped away to create an ion, all that is left is a bare proton, so proton and hydrogen ion are synonyms. These reactions are sometimes called a proton pump.

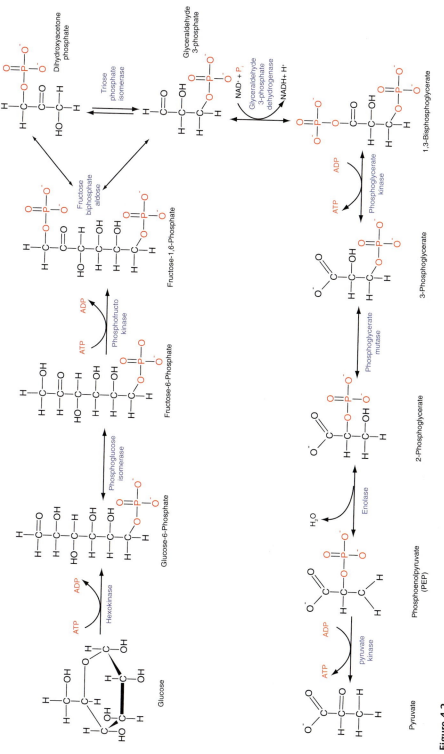

Figure 4.2

The glycolysis pathway. Starting with the glucose molecule at the upper left, and ending with pyruvate at the lower left, the figure shows the names and chemical structures of each substrate in the pathway. The straight arrows linking the structures identify the enzyme that catalyzes the reaction, and the curved arrows indicate other molecules that enter or exit the reaction. Phosphate groups are highlighted in red.

is shown in figure 4.2. The electron donor in this case is a molecule of the saccharide **glucose**, and the ultimate electron acceptor is a molecule called **pyruvate**. Consider the first reaction, which breaks the cyclic glucose into a straight-chain six-carbon sugar, and also adds a phosphate group; the resulting molecule is called glucose-6-phosphate. The arrow representing the reaction carries three important pieces of information. First, written in blue below the arrow, is the name of the enzyme that catalyzes this reaction: **hexokinase**. Enzymes that transfer phosphate groups are generically known as **kinases**; here the phosphate is attached to a six-carbon sugar, generically called a hexose (*hex* means six, and *ose* refers to a sugar), hence hexokinase. The process of attaching a phosphate group, called **phosphorylation**, plays many important roles throughout life. Second, note that the arrow has only one head, meaning that this is an irreversible reaction; many of the other reactions in this pathway are reversible, and hence have two-headed arrows. Third, the curved arrow attached to the reaction indicates that the reaction requires the consumption of an ATP, turning it into an ADP. There are two reasons the reaction needs ATP: It is the source of the phosphate group (seen in red in the figure) added to make glucose-6-phosphate, but more important, it provides the energy necessary to make this thermodynamically infeasible reaction possible. So far, this reaction has consumed, not produced, ATP, but keep following along.

The next step in glycolysis reconfigures the molecule so that a different carbon atom has a double bond with an oxygen atom. The glucose-6-phosphate reactant has a double-bonded carbon atom at the end of the molecule furthest from the phosphate; in the fructose-6-phosphate product, the double bond has been moved to another carbon. Since the molecular formulas for these two molecules are the same, they are isomers of each other; the name of the enzyme that catalyzes this reaction is **phosphoglucose isomerase**.[5] The next step again requires the input of energy, and again adds a phosphate group, this time to the other end of the molecule. The irreversible reaction is catalyzed by phosphofructokinase, and produces fructose-1,6-bisphosphate (a chemist's way of naming a fructose molecule with a phosphate group at each end). This reaction requires an ATP molecule, so rather than generating energy, this pathway has so far consumed two ATPs. The next step starts the process of getting more energy out than is going in, splitting the six-carbon sugar into two three-carbon molecules.

---

5. It is useful to decompose long enzyme names in order to guess what they do. The "phosphoglucose" part implies that the substrate of the enzyme is a glucose molecule with a phosphate group attached, and the "isomerase" suggests that the enzyme will produce an isomer of the substrate. Most enzymes have equally straightforward names.

One of the three-carbon molecules, glyceraldehyde-3-phosphate, is then converted into 1,3-bisphosphoglycerate through the incorporation of an inorganic phosphate atom. This is the first step in which free energy is produced. Here, the carrier of the energy is not ATP, but another molecule called **nicotinamide adenine dinucleotide**, but almost always written NAD+. A hydrogen atom and a free electron are transferred to this molecule in the reaction (the enzyme name of "dehydrogenase" suggests this), making NADH, which is the version of the molecule that can be made to do work. The addition of the electron changes the redox state of the molecule, and NADH is used when a reaction requires a change in redox potential (e.g., making a double bond). Not only is useful energy captured in NADH, but the stoichiometry means that two NADHs are produced, once for each of the two three-carbon molecules produced in the previous step.

The next step in the chain transfers that phosphate group back to an ADP, generating more useful energy as ATP. Since this happens twice (once for each of the three-carbon molecules), the pathway has now produced more useful energy than it has consumed. The final step produces another pair of ATP molecules, so the net gain is two ATP molecules and two NADH molecules for each glucose molecule going in. In addition, at the end there are two three-carbon molecules, called **pyruvate**, remaining.

Different organisms have different approaches to dealing with the resulting pyruvate: Those that can manage oxygen use it as an input to a much more efficient process for producing more energy, described in section 6.2.3. Others convert it into lactate or ethanol and excrete it as a waste product. It can also be converted into a variety of other substances needed by living things; for example, another pathway can convert pyruvate into fatty acids, the building blocks of cell membranes.

Glycolysis is actually one of the simplest metabolic pathways, involving relatively few inputs (glucose, ADP, NAD), few reactions, and few outputs (pyruvate, ATP, NADH). The glycolysis pathway functions to capture energy from food and bundle it up in packages that can be used to facilitate other metabolic reactions. Everything else the organism does also needs to be supported by a metabolic pathway. There are hundreds of pathways for synthesizing membranes, genomes, and other cell components, including making all the enzymes that catalyze the metabolic reactions.

Those biosynthesis pathways have to start from environmentally available inputs. Bacteria that lived in the early days of life didn't even have sugars to digest, since glucose was not naturally abundant (like oxygen, it's made mostly by living things). Some derived their energy from sources that were more environmentally abundant, like hydrogen sulfide from underwater volcanic vents (there are contemporary bacteria that still do this). Others used **photo-synthesis**, a chemical approach to capturing energy from sunlight, which is described in section 6.2.4. Photosynthesis became widespread fairly early in the history of life, unleashing rapid growth in the amount of both sugar and oxygen in the environment.

Glycolysis is also simple in that it connects to other pathways only at its inputs and outputs. More often, compounds in one metabolic pathway are also participants in another pathway, leading to complex branching and even cycles in the metabolic network. Defining exactly where one pathway ends and another begins is a matter of convenience; for the organism, the entire meta-bolic network is seamless, even though not all of it is necessarily functioning at the same time. All of the activities of living things that require transforming matter and energy, from digesting food to synthesizing DNA, are achieved by these complex, interrelated metabolic processes.

Mapping out all of the metabolic pathways and the relationships among their substrates for even such relatively simple organisms as bacteria is an enormous undertaking, involving many thousand substrates, enzymes, and reactions. The complete set of metabolic pathways, including all substrates, hasn't been completely enumerated for any single organism yet, although in certain well-studied organisms the vast majority are known. Interestingly enough, all of the enzymes are known for a large number of organisms. However, sometimes knowing the structure doesn't provide information about the function; for many of those enzymes, the substrates and activities remain to be discovered.

4.3   Chromosomes and Genomes

All of the information necessary to make a bacterium, that is, all of its heritable characteristics, forms its **genome**. Since bacteria reproduce asexually, their genomes are always haploid; recall from section 2.2, that means they have only one allele for every gene, not two. The structure that contains the genome is called a chromosome, the third major component of the cell. Bacteria gener-ally have a single, circular chromosome that contains the entire genome. Some bacteria have much smaller satellite bits of additional chromosome, called plasmids, which contain a small number of genes. Occasionally bacteria exchange plasmids, which can increase the rate at which advantageous genes are spread through bacterial populations.

Chromosomes have to support the two key functions required of a genome. First, the genome must be able to make high-fidelity copies of itself to pass onto offspring. Second, the genome must be able to embody all of the information necessary to make every structure that the organism might ever create. The activities supported by the genome are obviously central to reproductive success. Consider the issue of the speed at which a bacterium can reproduce. The step in bacterial reproduction that usually takes the most time is replicating the genome. A bacterium with a more compact genome will generally be able to reproduce faster than one with a larger genome. This fact makes clear the evolutionary pressure on bacteria to have the smallest possible genomes. Of course, there are other differences among bacteria that also matter for reproductive success; a larger genome can be balanced by a better ability to reproduce in a particular niche.

The molecular structure that supports those functions, the material that chromosomes are made of, is **deoxyribonucleic acid**, better known as DNA. DNA is a polymer, a long chain of repeating monomers called **nucleotides**. Four nucleotide bases are found in DNA: **guanine**, **cytosine**, **thymine** and **adenine**.[6] A fifth nucleic acid called **uracil** is found in RNA. The nucleotides are generally abbreviated G, C, T, A and U, respectively. The molecular structure of each nucleotide, shown in figure 4.3, contains three parts: a five-carbon sugar called **ribose** with one oxygen atom removed (hence **deoxyribose**), one of four different nucleoside **bases** (chemically, these are **dinitrogen heterocycles**, that is, carbon rings with two nitrogen atoms substituted for carbons), and then one or more phosphate groups, attached to the sugar. The phosphate group is attached to the sugar at the fifth carbon, called the 5′ (five prime) position. As shown in figure 4.4, the polymer is formed by linking the next nucleotide to that phosphate group. The linkage attaches to the next nucleotide at the third carbon in the sugar, called the 3′ position. These polymers can be extremely long: it is not at all unusual for a DNA molecule to contain millions, or even hundreds of millions, of bases.

Two important properties of this molecules are related to carrying out its coding and replication functions. First, note that it is possible to distinguish the ends of the molecule: one is called the 3′ end (where nothing is attached to the 3′ position of the last nucleotide) and the other the 5′ end (where there is nothing further attached to the 5′ phosphate). The fact that the two ends can be distinguished means that the sequence of bases has an unambiguous directionality, a beginning and an end. The precise sequence of nucleotides that makes up a DNA molecule is the encoding of all of the genetic information

---

6. We've already seen another function for adenine: When a sugar and three phosphate groups are attached, it becomes ATP. It is also part of the other energy-carrying molecule, NAD. Biological systems often use molecules, or modest chemical modifications to them, to achieve multiple functions. Once it is possible to metabolically synthesize something, adapting it to a new use can involve a small change in the genotype that can have a large effect on phenotype.

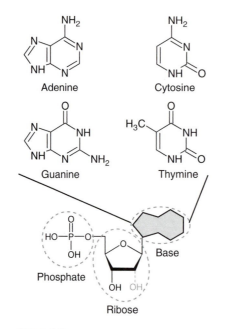

Figure 4.3
The chemical structure of nucleotides. Each nucleotide contains one of four nitrogen-containing cyclic bases (adenine, cytosine, guanine, or thymine) bonded to a five-carbon sugar with one oxygen atom removed (deoxyribose), which is in turn bonded to a phosphate group.

in the organism. There's a lot of information that goes into specifying even a bacterium with a relatively small genome. The DNA of the autotroph *Synechococcus elongates* contains 2,695,903 bases (human DNA is more than 1,000 times larger, containing more than 3 billion bases).

Second, as shown in figure 4.4, each base forms either two or three hydrogen bonds with one of the others, called its **complementary base**. T bonds with A, and G bonds with C. Since these hydrogen bonds are energetically favorable for the molecule, DNA is generally found in **double-stranded** form, where one polymer (or **strand**) is completely bonded to a **complementary strand**. Like a mirror image, the complementary strand inverts both the bases (substituting A's for all T's, T's for all A's, C's for all G's, and G's for all C's) and the direction (the 5′-to-3′ sequence of bases in one is complementary to the 3′-to-5′ sequence of bases in the other). The complementary bases in a double-stranded DNA molecule are called **base pairs**, and the size of DNA molecules is generally measured in base pairs, abbreviated bp (kbp is thousands of base pairs, mbp is millions of base pairs). Recall that hydrogen bonds are relatively weak, certainly compared to covalent ones, so it is possible to

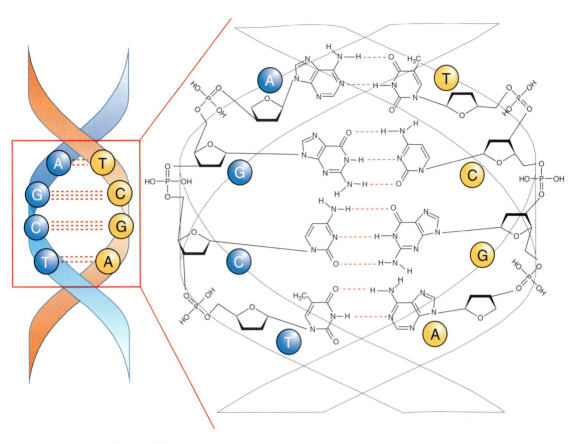

Figure 4.4
A schematic of how nucleotides are assembled into the double helix of DNA. Along the backbone of a single DNA strand, the phosphate groups link the 5′ carbon of one sugar to the 3′ carbon of the next. Two strands of DNA are attached to each other through hydrogen bonding between complementary nucleotides: adenine with thymine, or cytosine with guanine. Hydrogen bonds are shown as dotted red lines.

separate the two complementary strands of DNA by breaking the hydrogen bonds, without disrupting any of the covalent ones.

The complementary strands of DNA molecules make possible a straightforward and high-fidelity copying mechanism. To copy a DNA molecule, the complementary strands are first unzipped, by rupturing the hydrogen bonds. Then each of the single strands serves as a template for creating another DNA molecule that is complementary to it. When new strands with complementary base pairings to each of the original strands are synthesized, the result is two double-stranded DNA molecules that contain the exact same sequence of nucleotides as the original. This process is described in more detail in section 5.2.1.

The process of reproduction in bacteria is called **binary fission**, meaning one bacterium splits into two similarly sized offspring. The process begins with the duplication of the bacteria's chromosome. The chromosome copies then **segregate**, attaching to the cell membrane (sometimes via a group of proteins called a **mesosome**) at different places. The cell then elongates, pulling the chromosomes even further from each other. The membrane between the two separating chromosomes then invaginates, or pulls inwards, eventually splitting the cell into two parts, each of which contains about half of the prior cytoplasm and membrane from the parent, and one copy of the genome. All of this work is mediated by the concerted action of groups of proteins, powered by ATP and other energy sources. In addition to metabolic roles, proteins fulfill other important biological functions such as the physical movement of molecules.

Demonstrating that the sequence of DNA in the genome specifies all the information necessary for an organism to carry out all of its functions is relatively easy, say, by transplanting the chromosome of one organism into the cytoplasm of another. If it survives the experiment, the recipient changes character completely, taking on all the characteristics of the donor, including producing donor organism offspring. Understanding exactly how the structure of DNA encodes all of the heritable characteristics of an organism requires a bit more detail about enzymes and genomes, which are covered in chapter 5.

4.4   Bacterial Life

The essence of bacterial life is simple: Metabolize environmental energy and matter into biological molecules until there is enough material to divide; then repeat. However, long before even a few hundred doublings, there ceases to be enough environmental energy and matter to go around. Bacteria have to compete with each other, and some die. Rarely, advantageous variants arise, but when they do, they rapidly displace the previous champions. Faster reproduction is one way to gain advantage, but there are others as well; for example, being able to exploit a new energy source, or exploit an old one more efficiently. Even in the world of bacteria, it matters who the neighbors are. The environment, particularly the other organisms around, has a big impact on reproductive success. Bacteria tend to form communities, where the wastes of one bacterium become the valuable environmental inputs to another. Some bacteria became predators, feeding on other bacteria. Naturally, defenses against predation became adaptive shortly thereafter. Ecosystems arose, involving many sorts of organisms and very complex interdependencies.

Bacteria also dispersed throughout the environment. Different local circumstances (e.g., temperature, salinity, or pH) drive the evolution of variants that can out-compete their ancestors in a particular niche. Other innovations, like the flagellum, a whiplike appendage that many bacteria use to propel themselves, have proven generally useful. However, variations persist only if they result in gains of reproductive fitness that outweigh any cost entailed. The evolution of the flagellum is an illustrative case.

The flagellum is a long filament that rotates like a screw. It is driven by a molecular motor that uses the same flow of hydrogen ions that powers the ATP synthase discussed earlier, in this case to provide rotational energy. Not only is metabolic energy required to synthesize and transport the parts of this system, energy is required to run it. The very hydrogen ions that power the flagellum could have been used to generate ATP for other cellular needs. To improve reproductive success, a flagellum has to provide more energy (or another fitness payoff) than it costs to run it.

Powered movement pays off for the organism only when it can be controlled. Moving randomly is no different from being buffeted about the way bacteria without flagella are, and is unlikely to provide enough fitness value to be worth the price of propulsion. For a flagellum to be worth the required investment, it has to be used to move toward things that are of value (or away from ones that are harmful).

A flagellum is either on or off, though, and can't be steered. Nevertheless, flagella are tied directly to a system that accurately assesses the value of "here" versus the value of "somewhere else" from the bacterium's point of view. The entire molecular basis for this activity has been well worked out, and it involves a very modest number of proteins. The process, called **chemotaxis**, starts with protein molecules in the cell membrane sensing the environment for both food and toxins. These receptor molecules likely arose before flagella to modulate particular metabolic pathways; the molecular mechanisms that underlie this kind of regulation are discussed in section 5.2.1. The environmental signals also have to be integrated over time so as to respond to *changes* in the concentration of food and toxins; this is accomplished by enzymes that add and subtract methyl groups on the receptors when these substances are detected, accumulating information about trends over time. When the concentration of food has been falling for a while (or when the concentration of toxins has been rising), the flagellum rotates counterclockwise more frequently, which drives the bacterium forward. When the concentration of food has been rising for a while (or toxins falling), the flagellum rotates clockwise more often, which causes the bacterium to tumble in place, changing orientation at random. The flagellum changes from clockwise to counterclockwise and back frequently on its own, with or without external signals.

How does all this work? The bacterium still can't steer. However, the mathematics of random walks provides a strategy to get the bacterium to where it wants to go. When things are getting worse, change direction less often, moving away from where you are. When things are getting better, change direction more often, exploring the local area. Averaged over time, this control of the *frequency* of random changes in direction takes the bacterium toward what it likes and away from what it doesn't. This chemotaxis system, composed of just a few proteins, endows flagellar bacteria with the ability to, on average, swim up food gradients and down toxin gradients.

These bacteria have the basic ingredients of awareness, choice, and desire—realized in molecular structures. They are able to sense aspects of their environment, remember a bit of the past, and select among alternatives that influence their future, all in ways that are beneficial to their survival. Control of the flagellum is tied directly to an assessment of the value of the current environment and, implicitly, to the expected value of an alternative—the essence of what it means to have a desire. The substantial energetic cost of even this simple sort of sensation and action ensures that only organisms who can recoup that investment by creating better futures for themselves will survive. A selective advantage for internally representing information about the world and shaping action based on that representation appears frequently in the living world, reflected here in the broad distribution of bacterial flagella. It may not seem like much compared to human cognition, but chemotaxis genuinely involves representation and choice, based on what is of value to a bacterium. Since chemotaxis involves real costs to the organism, evolution ensures these representations and choices are meaningful, even in bacteria.[7]

4.5 Suggested Readings

Betsey Dexter Deyer's *A Field Guide to Bacteria* (Comstock, 2003) is a wonderfully written book that describes where to find and how to recognize dozens

---

7. The use of terms like *awareness*, *desire*, and *choice*, let alone *meaningful*, to describe the function of a set of molecules in a bacterium may seem like a stretch, but the activities of these molecules are clearly related to our own more cognitive versions of these activities. See Daniel Dennett's *Freedom Evolves* (Viking, 2003) for a philosophical approach to this issue, and Read Montague's *Why Choose This Book?* (Dutton, 2006) for the perspective from computational neuroscience. Both identify bacterial chemotaxis as having important commonalities with our hugely more elaborated systems for representing and valuing aspects of the world and choosing among possible actions based on those assessments.

of kinds of bacteria, without lab experience or equipment. Her technically careful, but accessible and entertaining book makes the bacterial world immediately tangible. Most microbiology books focus solely on disease-causing bacteria, but Jessica Snyder Sachs' *Good Germs, Bad Germs: Health and Survival in a Bacterial World* (Hill and Wang, 2007) is admirably balanced, touching on the critical roles played by bacteria not only in disease, but also in human health.

5 Biological Macromolecules

In bacteria, as in the rest of life, proteins and nucleic acids carry out most of the critical functions. Inheritance depends on DNA; metabolism relies on proteins. Proteins and nucleic acids, together called **macromolecules** because of their large size, are central to life. Their study is at the core of molecular biology.

Although individual molecules each have their own structure and functions, mostly proteins work together in groups. The proteins that catalyzed the many reactions in the glycolysis pathway that turns sugar into ATP are one example of how a group of proteins can work together in a pathway. Macromolecules can also physically assemble together into a **molecular machine**, a dynamic collection of many molecules that accomplishes a specific task. Many molecular machines are found in living things, including ones involved in protein production, energy metabolism, harnessing sunlight, packaging particular molecules for storage or destruction, and so forth. A molecular machine that involves a relatively small number of parts is sometimes called a **molecular complex**.

The metaphor of a machine is useful in thinking about these assemblies. Understanding a machine requires knowing what the parts are, and how they fit together in space and through time. Like any other, these molecular machines require energy to work, so it is important to specify how much power they consume and how they get it. Other important biological questions include how the machines are themselves constructed, how they get to the locations where they are used, how they are turned on and off or otherwise controlled, how they can fail, and how they are disposed of when broken or no longer necessary.

Evolution turns out to be a powerful tool in understanding molecular machines. Analysis of variants of a particular molecular machine seen in different organisms can provide insights into ancestral versions, and what modifications have accreted through time. Evolution also places a severe constraint on these mechanisms: At each stage of its evolution the machine had to function, and no new part could persist unless it immediately contributed somehow to fitness, usually through improved performance of the mechanism.

Before understanding the functioning of these molecular machines in detail, however, it is important to understand something about the structure and function of individual macromolecules.

5.1   Protein

Nearly all of the catalysts that drive metabolism are proteins. These catalysts are able to overcome activation barriers, and even make thermodynamically infeasible reactions occur. The chemical structure of proteins is what makes possible such diverse and remarkable activities.

Proteins are polymers, consisting of a linear sequence of components called **amino acids**. As shown in figure 5.1, each amino acid consists of three functional groups attached to a central carbon atom.[1] The central carbon and two of the groups remain the same in all amino acids; the third group varies, giving each somewhat different chemical properties.

The central carbon atom is called the **alpha carbon** (C_α). The NH_2 is an **amino** group (the name derives from ammonia, NH_3), and the group of $C=OOH$ is a carboxyl (see section 3.3), which makes the molecule an acid. The "R" in the figure indicates the variable group (historically called a radical, hence R), and each amino acid has a different one. The amino acids are strung together into a chain by linking the amino end of one to the carboxyl end of the next, like a stack of Lego™ blocks. As shown in figure 5.1, the linkage between the amino acids is called a **peptide bond**, and it involves bonding the nitrogen of the amino group with the carbon from the carboxyl group, freeing two hydrogen atoms and an oxygen (making water). Amino acids can be linked together this way indefinitely, forming chains called **polypeptides**.[2] The path through the polypeptide chain via the amino nitrogens, the alpha carbons, and the carboxyl carbons is called the **backbone**. The R groups for each amino acid hang off the backbone chain, and are called **side chains**. Since the peptide bond removes some of the atoms, the part that remains (most of the amino acid) is sometimes called an amino acid **residue**.

---

1. The hydrogen also attached to that carbon atom (see section 3.1.2) is also, technically speaking, a functional group.

2. Biologists tend to refer only to short chains as polypeptides; longer chains are generally called proteins. There's no strict boundary between the two, and strictly speaking, all proteins are polypeptides.

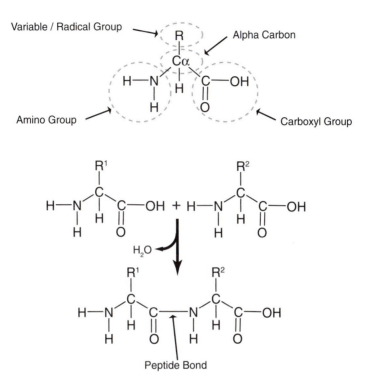

Figure 5.1

A schematic of amino acids, and how peptide bonds attach them. The upper portion of the figure shows the structure of an individual amino acid. It consists of four chemical groups: a nitrogen-containing amino group, a carboxyl group, a central or "alpha" carbon, and a variable group, which varies among the different amino acids. The lower portion of the figure shows how peptide bonds are formed. The OH from the carboxyl group of one amino acid reacts with the H from the amino group of the next amino acid, releasing water, and forming a peptide bond between the two amino acids.

Twenty different amino acids make up the proteins of most organisms, and each has somewhat different chemical properties from the others. For example, the simplest possible R group is just a single hydrogen atom; the amino acid made that way is called **glycine**. **Alanine**'s side chain is a similarly simple CH_3; **tryptophan** has the largest side chain consisting of nine carbons, eight hydrogens, and a nitrogen atom in two linked rings. In addition to differing in size, side chain groups confer important chemical properties. **Leucine** is hydrophobic, interacting well with lipids. **Serine** is polar, preferring to interact with water. **Histidine** ionizes easily, becoming charged. **Cysteine** contains a sulfur atom, conferring other chemical properties. Table 5.1 shows some of the basic features of each amino acid.

Table 5.1
Some of the properties of the 20 amino acids that make up most proteins

Amino Acid	Abbrev (1)	Abbrev (3)	Mass	Side Chain	Hydrophobic	pH	van der Waals Radius
Alanine	A	Ala	89.09	-CH3	yes	—	67
Cysteine	C	Cys	121.15	-CH2SH	yes	acidic	86
Aspartic acid	D	Asp	133.10	-CH2COOH	no	acidic	91
Glutamic acid	E	Glu	147.13	-CH2CH2COOH	yes	acidic	109
Phenylalanine	F	Phe	165.19	-CH2C6H5	yes	—	135
Glycine	G	Gly	75.07	-H	no	—	48
Histidine	H	His	155.16	-CH2-C3H3N2	no	weak basic	118
Isoleucine	I	Ile	131.17	-CH(CH3)CH2CH3	yes	—	124
Lysine	K	Lys	146.19	-(CH2)4NH2	no	basic	135
Leucine	L	Leu	131.17	-CH2CH(CH3)2	yes	—	124
Methionine	M	Met	149.21	-CH2CH2SCH3	yes	—	124
Asparagine	N	Asn	132.12	-CH2CONH2	no	—	96
Proline	P	Pro	115.13	-CH2CH2CH2-	yes	—	90
Glutamine	Q	Gln	146.15	-CH2CH2CONH2	no	—	114
Arginine	R	Arg	174.20	-(CH2)3NH-C(NH)NH2	no	strongly basic	148
Serine	S	Ser	105.09	-CH2OH	no	—	73
Threonine	T	Thr	119.12	-CH(OH)CH3	no	weak acidic	93
Valine	V	Val	117.15	-CH(CH3)2	yes	—	105
Tryptophan	W	Trp	204.23	-CH2C8H6N	yes	—	163
Tyrosine	Y	Tyr	181.19	-CH2-C6H4OH	no	—	141

Note: Each row shows the name, one- and three-letter abbreviations, the mass, the chemical structure of the variable region, whether or not that variable region is hydrophobic, what the pH of the amino acid is, and its van der Waals radius, which indicates the size of the amino acid.

Composing these various elements in a specific order can create proteins with an enormous variety of chemical properties. Each protein is defined by its **amino acid sequence**, which is just the ordered list of amino acids reading down the backbone from the end with an unbound amino group (the **N terminus**) to the end with an unbound carboxyl group (the **C terminus**). Exactly what chemical properties the protein will have as a whole depends on the particular shape that sequence takes on.

Some parts of a polypeptide chain are flexible. Most bonds have a narrow range of angles that are energetically favorable, so the shapes of molecules containing them are effectively fixed. However, there are three places in a polypeptide chain where the bond angles are free to rotate. Most important, the two dihedral angles (angles between planes) between adjacent amino acids,

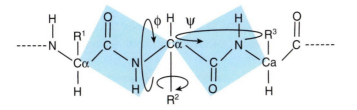

Figure 5.2
A schematic showing the three places in a polypeptide where the bonds are free to rotate. The variable region of each amino acid (shown by the R) is free to rotate around its axis. The planes formed by the rigid parts of the amino acid can rotate with respect to either of two axes (called phi and psi dihedral angles).

called the ϕ (phi) and φ (psi) angles, can rotate freely, as shown in figure 5.2. A polypeptide really is like a chain, where the individual amino acids are rigid, but the linkages between them are quite flexible.

The third rotatable bond is the one that links the side chain to the central carbon atom in each amino acid. That means the orientation of the side chain with respect to the backbone is also flexible. It's as if each link in the chain had an elaborate bead (the side chain) attached that could spin around its axis, but not move otherwise.

The flexibility of the backbone means that a polypeptide chain can take on a very large number of possible shapes, called **conformations**. Short polypeptides are usually floppy, rapidly jumping around through the large space of nearly equally probable shapes. However, in longer chains the different conformations can have substantially different energies; proteins are mostly found in a single, lowest energy state, which is called the **three-dimensional structure**, or sometimes the **fold**, of the protein. It is this particular shape of a protein that gives an enzyme its remarkable chemical properties. The shape of a protein is described by specifying values for all of the rotatable angles, or equivalently, the positions of all of the atoms in the molecule.

5.1.1 Protein Structure

It is relatively simple to demonstrate that most proteins have a single preferred conformation in the conditions associated with life (e.g., temperature and pH). Heating a protein, for example, will often cause it to **denature**, and take on a different conformation, changing its function.[3] Dehydration denatures

---

3. Many of the activities of cooking, such as heating and whipping, change the structure of the proteins in food. See, for example, Harold McGee's *On food and cooking* (Scribner, 2004). Detergents also denature proteins, making them easier to wash off.

proteins reversibly; most denatured proteins will spontaneously refold into their native conformation in water, demonstrating that the protein's amino acid sequence alone is generally enough to determine its three-dimensional structure[4].

Most proteins reside entirely in aqueous solution and take on a compact, roughly ball-like shape. These proteins are called **globular**. One of the major forces driving the conformation of globular proteins is the tendency of hydrophobic side chains to be buried inside the protein (therefore shielded from water), and the polar side chains to be exposed on the surface (so they can interact with water). Not all proteins are globular; for example, the ones that are bound in membranes, like those that form the pores and channels described in section 4.1, are at least partially protected from water, and therefore shaped by different forces.

The amino acids in a protein can also interact with each other, forming hydrogen bonds or other types of stabilizing connections that affect the overall conformation. Some of those interactions are between amino acids that are linked by a peptide bond, or are at least close to each other in the backbone chain. However, some of the interactions are long distance, that is, between amino acid residues that are far from each other in the chain. The forces that drive a particular amino acid sequence to assume its particular conformation are so complex that, though it is simple to write down the equations, actually calculating the minimum energy conformation is well beyond the abilities of even the most powerful computers. This calculation, called the **protein folding problem** or sometimes **protein structure prediction**, is one of the great open computational challenges of molecular biology. In the meantime, protein structures are experimentally determined by the techniques of X-ray crystallography and nuclear magnetic resonance, described in section 10.1.2.

Protein structures have very detailed landscapes, with many protrusions and pockets. Figure 5.3 shows one of the smallest and simplest proteins, **insulin**. Note the complex external shape, and the fact that the hydrophobic residues (colored in red) are buried and the polar residues (blue) are exposed to the water (see figure 1.6b for another view of the same molecule). Like many other proteins, insulin forms **dimers** in which two identical molecules pair with each other to form the active compound; the figure shows the linked pair of insulin chains. Figure 5.4 shows the structure of a much larger (the

---

4. A few proteins do not fold spontaneously into their biologically active shape; other proteins called chaperonins are required to ensure that they fold properly. Conversely, prions are disease-causing proteins that can cause normally folded proteins to change into an incorrect shape. Mad cow disease is caused by prions.

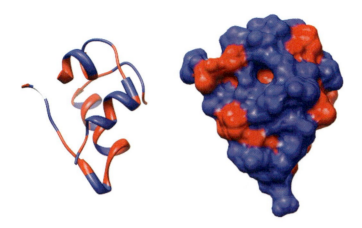

Figure 5.3
Two molecular renderings of the small protein insulin. On the left, a ribbon diagram, illustrating the helices and turns (there are no sheets in insulin). On the right is a space-filling rendition, showing the complex external shape. Hydrophobic residues are colored red, and polar residues are colored blue.

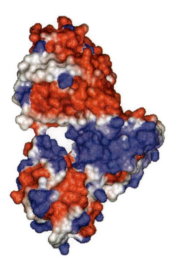

Figure 5.4
A space-filling rendition of a large bacterial protein called the Tu elongation factor. Here, red indicates positively charged regions, blue indicates negatively charged ones, and white neutral ones. Adapted from Figure 4 of Das, et al., *BMC Genomics* 7 (2006), 186 (doi:10.1186/1471-2164-7-186) and used under the Creative Commons Attribution License.

figures are not on the same scale) and more complex protein from a bacterium, rendered so that only the surface is visible, colored to indicate the distribution of charge.

A protein's folded structure determines its function. For enzymes, at least a small patch of the protein's surface, called a **recognition site**, fits the molecule that it acts on (its substrate) very precisely.[5] The exact fit between the shape of the protein and the shape of the substrate, and often a similarly exact complementarity in the distribution of charges between the protein surface and the substrate, underlie the amazing specificity of protein catalysts. Complementarity of shape and charge are how proteins manage to recognize and act on only their very specific of targets.

Enzymes also have **active sites** where substrates that have been recognized and bound to the protein are acted on, say by breaking a bond. Active sites are often relatively inaccessible, so that only substrates that fit exactly into a specifically shaped cavity are processed. The shape and charge distribution on the surface of a protein can also be arranged to bring two substrates close together and hold them there long enough to facilitate a reaction between them.

Protein structures also make possible other functions. For example, regulatory proteins turn on and off (or modulate) the activity of other proteins. In one mechanism, called **allosteric** regulation, the regulator protein binds to a site on the regulated protein that indirectly affects the target's active or recognition sites. Proteins that accomplish nonenzymatic functions (e.g., ones that act as channels or transport other molecules) also gain their activity through their conformations.

Since protein structures are so complicated and so central, biologists use various ways of describing them that emphasize different important aspects. The sequence of amino acids itself, often represented as a string of "letters" (each character standing for one of the 20 amino acids; see table 5.1), is the most fundamental and is sometimes called the protein's **primary structure**, or more frequently its **sequence**.

That primary sequence folds up into more complex structures. The backbone of the sequence follows a complicated path in a folded protein, but the kinds of paths observed in nature can be broken down into a few broad classes. The backbone crosses back and forth many times through a globular protein, and the amino acids at those transitions form structures called **turns**. A path can be stretched out so that adjacent amino acids are about as

---

5. Enzymes are usually much, much larger than their substrates.

far apart as they can be; this is called a **beta strand**. Two or more strands that touch each other, forming hydrogen bonds, make a **beta sheet**. At least two strands are necessary to stabilize this shape, so strands never appear except in a sheet. If the strands go in opposite directions, as they will when a strand makes a **hairpin turn** and then doubles back through the protein, it is called an **antiparallel beta sheet**. In contrast, when amino acids are packed as closely together as possible, they take on corkscrew-like shapes called **alpha helices**. Helices are stabilized by hydrogen bonds across the turns of the corkscrew.

Most of the backbone in most proteins is generally found in helix, sheet, or turn shapes. The composition of a protein in these terms is called its **secondary structure**, which is a useful level for seeing aspects of the protein's structure without being overwhelmed by the atomic details. Often protein structures are drawn as cartoons representing only idealized secondary structure elements, as in figure 5.5. Another useful distinction that can be seen in such cartoons is between the tightly packed central protein **core**, and the somewhat more flexible external **loops**, the backbone turns that reach the surface of the protein. The broad division into helix, sheet and turn is quite coarse, and other more detailed breakdowns of subclasses of protein structural components have been proposed, but secondary structure characterizations remain popular and useful.

The information about the precise position of every atom (or, equivalently, every phi/psi and sidechain angle) is called the **tertiary structure** of the protein, which is what figures 5.3 and 5.4 illustrate. And the way the protein fits together with either other instances of itself (like the insulin dimer), with other proteins, or sometimes with other small molecules (e.g., iron in the blood protein hemoglobin) is called its **quaternary structure**. Figure 5.6 shows the primary, secondary, tertiary, and quaternary structures of a protein. Biologists use all of these different ways of looking at structure to understand how the remarkable chemical properties of proteins arise. Analysis of structure is also useful for inferring what a protein of unknown function might be doing in a cell.

There are also a number of changes that can be made to a protein after it is synthesized, called **posttranslational modifications** (abbreviated PTM). Usually these modifications result in changes to the protein's activities, providing an important mechanism of control. For example, sometimes several proteins are synthesized as a unit (called a **polyprotein**) and become active only when they are cleaved. This is one way to prevent a potentially destructive protein (say, one that attacks invaders) from becoming active before it can be transported to the proper cellular context. Another important PTM is the

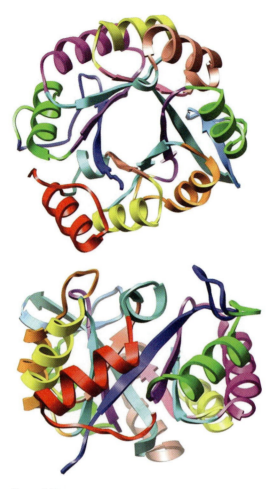

Figure 5.5
A secondary structure cartoon rendition of a protein called a TIM barrel, viewed at two different angles. The coloring traces the backbone starting in blue at the amino terminus ending in red at the carboxy terminus. The corkscrews indicate alpha helical structure, the arrows beta sheet.

Primary Structure

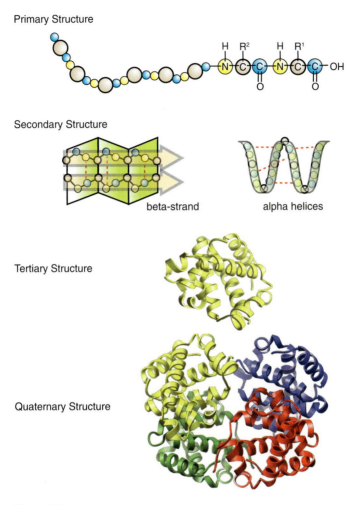

Secondary Structure

beta-strand   alpha helices

Tertiary Structure

Quaternary Structure

Figure 5.6
Cartoon renditions of the different levels of protein structure. The top panel illustrates the amino acid sequence, or primary structure of a protein. The next row shows how local relations among amino acids near each other can form sheets or helices, collectively known as the secondary structure of the protein (the red dotted lines indicate hydrogen bonds). The tertiary structure shows how the local secondary structures are packed together to form the complete three-dimensional structure of the protein. The quaternary structure, shown at the bottom of the figure, can involve multiple copies of a protein that assemble together into the active form (here, color indicates four different copies of the same protein assembling together).

covalent attachment of a chemical group (e.g., a phosphate) to a particular amino acid. Many proteins have to respond quickly to the state of the cell or the environment; posttranslational changes can happen much more quickly than the synthesis of a whole new protein. For example, membrane-bound receptors generally deliver the information that they have detected something through a series of changes in the PTM status of other proteins that act as messengers (detailed in section 7.2).

5.1.2 Protein Evolution

There are an enormous number of possible proteins,[6] so only a tiny fraction are ever observed in actual organisms. Interestingly, many living things have proteins that are quite similar to each other. For example, the insulin protein found in pigs differs from that found in humans by only one amino acid at the very end. The insulin found in cows differs by only three amino acids. The explanation is, of course, that all of these proteins descended from a common ancestor. Characteristics of organisms that are similar to each other due to common descent are called homologs; so, for example, pig, cow, and human insulin are homologs. Some of these homologies span enormous evolutionary time periods; quite a few human proteins have homologs in bacteria.

Generally speaking, a protein's amino acid sequence determines its structure, and its structure determines its function. Proteins with similar sequences tend to have similar structures, and proteins with similar structures (regardless of sequence) tend to have similar functions. This observation makes the ability to identify homologous proteins in different organisms very valuable. Understanding the structure or function of a protein in one organism gives pretty reliable information about the structure and function of its homologs in other organisms. As discussed in section 10.1.5, it's much easier to determine the sequence of a protein experimentally than it is to determine its structure or function, so reasoning by homology is very important in contemporary molecular biology.

Since the millions of different proteins that have been sequenced are descendants of various common ancestors, it is productive to group them into families. There are various ways to group together protein structures,[7] but they all

6. There are far more possible proteins of 1,000 amino acids or fewer than there are atoms in the observable universe. In fact, the number of possible proteins of exactly 100 amino acids (20^{100} or more than 10^{130}) is hugely greater than the number of atoms in the universe (about 10^{80}).

7. The two most widely used clusterings are SCOP (http://scop.mrc-lmb.cam.ac.uk) and CATH (http://www.cathdb.info/).

share broad outlines. The most basic structural classes are proteins with mainly helical secondary structures, those that have mainly sheets, those with a mixture of both helices and sheets, and those with relatively little helical or sheet content at all. Within those broad classes are subgroups defined by common structural motifs, such as the TIM barrel shown in figure 5.5. By further specifying the particular connectivity of the secondary structural elements along the protein's backbone, these broad classes are divided into fold families. At the most fine-grained level, groups of proteins called **super-families** are defined not only by similar structures, but also by clear homology to each other. The structural class of a protein provides important information regarding its function and evolution.

5.1.3 Protein Function

In addition to the structural classifications, a hierarchy of functional classes of enzymatic proteins is maintained by an international group of biochemists.[8] There are six top-level classes, and hundreds of subclasses. These classifications are useful in organizing the thousands of different sorts of reactions that enzymes catalyze into groups that share important properties. The six top level classes are (1) **oxidoreductases**, which catalyze redox reactions, mostly by transferring hydrogen atoms, so many of the individual enzymes are called **dehydrogenases**; (2) **transferases**, which transfer a chemical group from one compound to another, such as the kinases, which transfer phosphate groups or the DNA-directed **polymerases**; (3) **hydrolases**, which catalyze reactions that involve water, like the **peptidase** that breaks a peptide bond; (4) **lyases**, which break or create double bonds, like the **carboxylases** that work on C=OOH groups; (5) **isomerases**, which catalyze changes within a single molecule, such as enzymes that transfer a chemical group from one part of a molecule to another; and finally (6) the **ligases**, which join two large molecules using power from ATP, such as the enzymes that attach amino acids to the proper transfer RNA (see below).

Proteins are the instruments that make metabolism happen. Their functions include all of the enzymatic activities required for metabolism. Every metabolic reaction requires a protein (sometimes more than one). Nearly every step of every transformation along the path of taking environmentally available matter and energy and transforming it into self and offspring is mediated by a protein. It is not much of an exaggeration to say that the amino acid sequences of all of an organism's proteins tells you almost everything you

8. See http://www.chem.qmul.ac.uk/iubmb/enzyme/.

need to know about how the organism functions. In fact, that's exactly what its genome encodes.

5.2 Nucleic Acids

The proteins an organism can produce define the metabolic pathways the organism can use to turn available matter and energy into offspring. Proteins also have the particulate quality, discussed in section 2.1, of Mendel's units of inheritance; each protein could be inherited from one parent or the other, without blending. Furthermore, most small variations in the sequence of a protein might not generally do much (like the difference between pig and human insulin), but some small changes might affect, say, the active site, or the ability to recognize its target substrate or other major aspects of the protein's function. These changes in proteins seemed like a good candidate for the mutations that underlie Darwin's account of variation. Despite the prevailing scientific wisdom during the first half of the twentieth century, genes are not themselves proteins. Overwhelming evidence demonstrates that chromosomes, which are made of DNA, are the carriers of an organism's inheritance.

Instead, it turns out that genes are *about* proteins. A gene is a set of instructions about how (and when) to make a protein. Making proteins is such an important aspect of metabolism that there is a specific molecular machine, called the **ribosome**, which is responsible for producing all of the proteins in a cell. The ribosome is able to make proteins of any sequence from available pools of unbonded amino acids. Chemically, making peptide bonds between the amino acids is not a complex reaction. However, reliably producing the exact amino acid sequence of a particular protein is a much more difficult task. Somewhere there must be a mechanism that directs the protein synthesis machinery regarding which amino acids need to be attached to each other and in what order. *That* is the role of the gene.

5.2.1 DNA

How can something made of DNA direct the synthesis of a protein? That question has two answers, one structural and one functional. Functionally, some sequence of nucleotides in DNA corresponds to the sequence of amino acids for every protein. A gene contains a DNA sequence that specifies a protein; biologists say the gene *codes for* the protein. The DNA in a chromosome is very long, and can contain the sequences of thousands of different genes. Only some of the DNA in a chromosome specifies the sequence of a protein. The parts that do specify a protein are called **coding sequences** (CDS),

and the parts that do not are called noncoding. The division of DNA into protein-coding versus noncoding regions is one of the most fundamental distinctions in molecular biology.

A central molecular activity of all cells is the production of protein from the coding regions of its DNA. The process of turning the information contained in the DNA sequence into an active protein is called the **expression** of a gene. Noncoding **regulatory sequences** of DNA near a coding region help determine under what circumstances the gene is expressed, a process described in more detail in the following and section 6.1.

There is a direct translation between the sequence of the DNA that makes up a gene and the amino acid sequence of a protein. Since proteins have a larger number of monomeric building blocks (20 amino acids) than DNA (4 nucleotides), the mapping between DNA sequence and amino acid sequence can't be one-to-one. It takes three consecutive nucleotides (together called a **codon**) to specify one amino acid.

The specific mapping from codons to amino acids is called the **genetic code**. There are 64 different codons (because there are 64 three-letter combinations that can be created with a four-letter alphabet), but only 20 amino acids, so there is some redundancy in the mappings. For example, two codons, AAA and AAG (where A means adenine and G means guanine in the DNA sequence) both code for the amino acid lysine. Three special codons—TAG, TGA, and TAA—called **stop codons**, indicate the end of the coding sequence. The mapping between codons and amino acids has some minor variants in different organisms, particularly bacteria, but is largely conserved throughout life.

The structural answer to the question of how a DNA molecule directs the synthesis of proteins is so important to the understanding of life that biologists call it the **central dogma of molecular biology**. All living things, from bacteria to people, share this most basic process, which is impressive evidence for a universal common ancestor of all creatures.[9] Briefly stated in section 1.3, the central dogma holds that protein-coding regions of DNA are **transcribed** into RNA molecules, which are then transported to the ribosome and **translated** into protein. DNA sequence determines RNA sequence, which in turn determines protein sequence.

Not every protein is needed at the same time or in the same amounts, so a cell needs a mechanism to determine how much of which proteins are expressed at any given time. An important part of this mechanism is **transcriptional**

---

9. A few organisms (e.g., the HIV virus) have evolved unusual ways to exploit the process, and are in some ways exceptions to the central dogma. However, these organisms clearly arose from the same ancestors as the rest.

control, the determination when a gene will be transcribed into messenger RNA. Transcriptional control starts with a set of proteins, called **transcription factors**, that are needed to begin the process of turning a gene into a protein. Different genes are controlled by different transcriptions factors. A gene is transcribed only when its specific transcription factors are present and are interacting with the DNA. Transcription factors work by chemically recognizing and binding to specific DNA sequences, called **promoters**, which are found just before the coding region of a gene. Promoter sequences are often shared among proteins that are needed together (say, the set of proteins that make up a pathway), so one transcription factor can turn on the production of many genes. Some genes or groups require multiple transcription factors in order to be expressed, providing a mechanism that can create the nuanced responses a cell needs in different circumstances.

Cells use all kinds of complicated relationships among multiple transcription factors to regulate the expression of their genes. One illustrative (and relatively simple!) example is how the bacterium *Escherichia coli* regulates the expression of digestive enzymes based on the availability of different energy sources. These bacteria prefer to consume the sugar glucose. When glucose is not environmentally available but lactose (milk sugar) is, it switches its metabolism over to digest lactose. The switch involves turning on a collection of genes for enzymes that process lactose and turn it into glucose. (This process takes energy, which is why the bacteria prefer glucose.) The genes for the lactose metabolism pathway are always present, but are usually turned off (not expressed) by an inhibitory transcription factor called a **repressor**. This repressor binds to the DNA before the lactose genes, blocking access to the promoters and preventing their expression. In addition to binding DNA, a different part of this repressor protein also binds lactose. When lactose is present and binds to the repressor, that changes the structure of the DNA-binding part of the protein, causing it to fall off the DNA. Without the repressor blocking access of other transcription factors to the promoters, the genes for the lactose-digesting enzymes start being expressed. In bacteria and some other organisms, groups of functionally related genes (like those in the lactose metabolism pathway) are found adjacent to each other in the genome. The expression of these groupings of genes, called **operons**, are controlled by a single set of transcription factors, ensuring that they are all produced together.

Once the right set of transcription factors have bound to the promoter region near the beginning of a gene, the process of transcription begins with the attraction of a critical molecular complex, called **RNA polymerase II**. This polymerase binds to a DNA sequence next to the promoter that indicates the beginning of a gene or operon. Once the polymerase has bound, it recruits additional enzymes that unwind the DNA and use it as a template for the production of **messenger RNA** (mRNA).

5.2.2 RNA

RNA is a nucleic acid like DNA, but made with a slightly different backbone and a slightly different set of monomers: RNA uses a nucleotide called uracil (U) where DNA would use a thymine (T). RNA is also different from DNA in that it is mostly found in single-stranded form. The nucleotides in RNA sequences are complementary to those in DNA sequences, meaning that they obey the same base-pairing rules that underlie the ability to copy DNA sequences (described in section 4.3), with U being complementary to A. Those base-pairing rules mean that the messenger RNA molecule created by the polymerase captures the information in the DNA sequence of the gene. For example, if the gene contained the sequence GTAC, it would cause the polymerase to create a molecule of RNA with the complementary sequence CAUG. When the polymerase encounters a stop codon, it falls off the DNA, having directed the synthesis of an mRNA molecule that contains the complementary sequence for a complete coding sequence. That is the process of transcription; the resulting mRNA molecules are sometimes called **transcripts**.

The activity of many enzymes and a large amount of ATP is required for the process of transcription. RNA polymerase II is a kind of **DNA-directed RNA polymerase**. Decode that rather ungainly name by working backwards: It is an enzyme (the—*ase* ending indicates enzymes), it adds monomeric units to extend a polymer (a polymerase), the polymer it extends is RNA (an RNA polymerase), and the specific nucleotide appended is determined by reading from a DNA molecule (DNA-directed). RNA polymerase II recruits many other molecules (at least 10 more proteins in bacteria, and far more than that in some organisms) that play a role in properly processing the DNA to produce a transcript.

The completed RNA transcript of a gene is transported to a **ribosome**, where it is translated into protein. Ribosomes themselves are complex assemblies of proteins and RNA molecules. Bacterial ribosomes (called 70S ribosomes, after their separation band in an instrument called a centrifuge) are made of 3 different RNA molecules and 55 proteins. RNA has many roles in living things, so biologists often use a lowercase prefix to indicate what kind of RNA is being talked about. The RNA molecules that carry a transcript are called messenger RNA, written mRNA; **ribosomal RNAs** are indicated as **rRNA**.

A third kind of RNA, called **transfer RNA** (**tRNA**), brings the necessary amino acids to the ribosome, and functions to match nucleic acid codons with

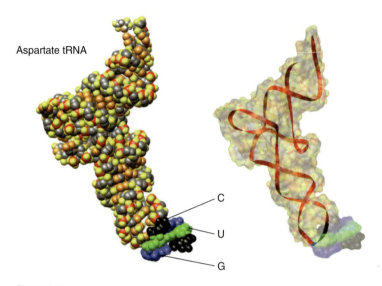

Aspartate tRNA

C

U

G

Figure 5.7
Two renditions of the transfer RNA for the amino acid aspartate. On the left, a space-filling rendition, showing the anticodon at the bottom, and the amino acid–binding region at the top. The rendition on the right shows the backbone of the RNA molecule that gives the tRNA its structure.

amino acids. There are many different tRNAs; at least one is needed for each of the 61 codons that specify an amino acid. The three-dimensional structure of a tRNA, shown in figure 5.7, includes a region at one end that binds to the appropriate amino acid, and, at the other end, a sequence, called an anticodon, that is complementary to a codon in mRNA.

There are, of course, genes for tRNAs and rRNAs in the genome. RNA coding genes are transcribed by a different polymerase, called **RNA polymerase III**, and are not processed by a ribosome, since no protein need be produced. In addition, tRNA molecules have to be linked to the proper amino acid, a task accomplished by enzymes with the clunky, but perhaps by now understandable name **aminoacyl-tRNA synthetase** (it's an enzyme that ligates, or binds chemically, a tRNA with its matching amino acid). Specific synthetases exist for each different tRNA.

All of these different parts come together at the ribosome, where the translation of gene into protein occurs, as illustrated in figure 5.8. Initially, a set of enzymes called **initiation factors** lock the 5′ end of the mRNA in place on the ribosome, and recruit a tRNA with the anticodon matching the first codon of the mRNA. Once the first amino acid is in place, enzymes called **elongation factors** move the ribosome down three nucleotides on the mRNA, and the

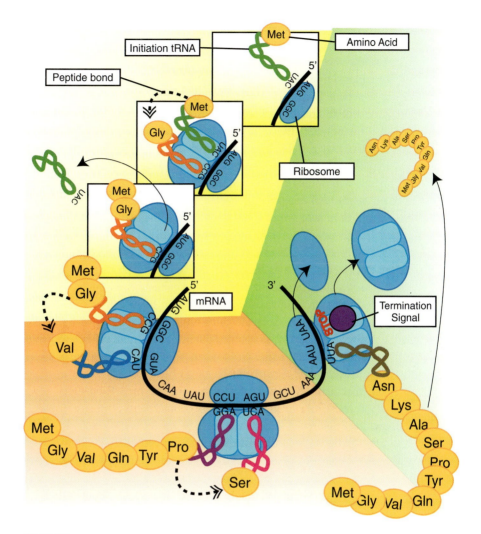

Figure 5.8
A cartoon of the translational apparatus at the ribosome. Starting at the top and moving counterclockwise: The process begins when initiation factors attach first the small ribosomal subunit and then the large subunit to the 5′ end of a messenger RNA (shown as a black line), and recruit the first tRNA (usually a methionine). In elongation, additional tRNAs are recruiting to match the sequence of the mRNA, and peptide bonds (dotted arrows) are formed between adjacent amino acids. After peptide bonds are formed, the ribosome moves down the mRNA, the tRNA is cleaved off, and the process is repeated. When the ribosome encounters a stop codon in the mRNA, termination factors cause the now complete peptide chain to be released from the ribosome.

tRNA that matches the second codon attaches. An enzyme called an **amino acid polymerase** catalyzes the reaction that creates a peptide bond between the new amino acid and the previous one, and then cleaves off the tRNA. The elongation process of moving a step, recruiting the next tRNA, and making a peptide bond then repeats until the end of the mRNA, when a complete polypeptide chain has been synthesized. A set of enzymes, called **termination factors**, then frees the polypeptide chain from the ribosome, and readies it for the next mRNA. The freed protein folds up, assuming its minimum energy conformation, and becomes ready for use.

Ribosomes are an excellent example of the molecular machines described at the beginning of this chapter. These machines involve the combined activity of dozens of proteins and RNAs to reliably accomplish one of the most central tasks of a living thing: protein synthesis.

Since the statement of the central dogma in the 1950s, the focus of much molecular biology has been on DNA and proteins. However, recently the roles of various kinds of RNA have gained increasing prominence. One reason for this change was the discovery that RNA can act as a catalyst. A single-stranded RNA molecule can form internal bonds among its complementary nucleotides (e.g., as visible in the structure of the tRNA shown in figure 5.7), meaning that some RNAs can fold into well-defined three-dimensional structures. These structures can give RNA enzymes (called **ribozymes**) very specific activities, much like protein enzymes. Ribozymes now play a fairly modest role in most organisms compared to protein catalysts, although the functioning of the ribosome is based on RNA catalysis, with the proteins playing a supporting role. The macromolecules of the ribosome in humans and bacteria are homologous, showing evidence that ribosomes existed at the time of our last common ancestor nearly 2 billion years ago. A plausible, but controversial and certainly unproven hypothesis about early life forms, called the **RNA world**, holds that RNA molecules predated both DNA and protein, playing both catalytic and information storage roles at one stage in the history of life.

A second reason for the resurgence of interest in RNA is the phenomenon called **RNA interference**, abbreviated RNAi. It had been observed for some time that messenger RNA molecules sometimes don't get translated into protein. Recently, the mechanism of this silencing was discovered, and it involves double-stranded RNA—a molecule not previously thought to have a role in normal metabolism. When a double-stranded RNA that matches the sequence of a normal, single-stranded mRNA is present in the cytoplasm, a molecular machine called the **RNA-induced silencing complex** chews up the single-stranded mRNA before it can be translated.

Evolutionary analysis suggests that RNAi mechanism started as a mechanism that bacteria used to protect themselves against viruses (some of whose genomes are double-stranded RNA). Most organisms have some sort of RNAi system, although it has evolved beyond its antiviral role into a method for controlling the expression of a large number of genes at once.

Organisms can also produce their own double-stranded RNAs from short pieces of RNA, called **micro-RNAs** (written **miRNA**). Since miRNAs are short sequences, they can match and therefore silence mRNAs from many different genes. Recent evidence suggests that a single miRNA can match hundreds of genes and therefore shut them all off simultaneously. Many of the proteins that are affected by miRNAs are transcription factors, which themselves regulate the expression of many more genes, suggesting that RNAi is a master regulator, turning on and off very large groups of genes. This is particularly important during development of multicellular organisms and in the differentiation of cell fates, described in section 7.2.4.

In addition to RNAi, there is a growing body of evidence that other RNA transcripts that do not code for proteins, called **noncoding** or **ncRNA**, play an important, although yet undetermined role in the cell. In human beings, more than 80% of DNA is transcribed, even though only about 5% is ultimately translated into protein. Although the function, if any, of all this ncRNA remains unknown, there is growing evidence of differential, tissue-specific differences in its expression. The once popular idea that a large portion of human DNA is nonfunctional "junk" is slowly yielding to an acceptance that there is much more to be learned.[10] The term **gene product** is sometimes used to be agnostic about exactly what functional product (protein or RNA) is produced at the direction of the DNA.

5.2.3   DNA Evolution

DNA directs the synthesis of proteins and RNAs, and therefore all the metabolic processes that transform available matter and energy into self and offspring. DNA also provides the mechanism by which these traits are passed on to offspring, so any heritable variation must involve a change in DNA.

As described in section 4.3, the duplication of an organism's genome is realized by the creation of an exact copy of its DNA. Figure 5.9 illustrates the detailed mechanism behind this process. An enzyme called a **topoisomerase** unwinds a bit of the DNA double helix, allowing another enzyme, called a

10. See for example, Elizabeth Pennisi's story in the journal *Science*, DNA study forces rethink of what it means to be a gene, 316 (2007), 1556–1557.

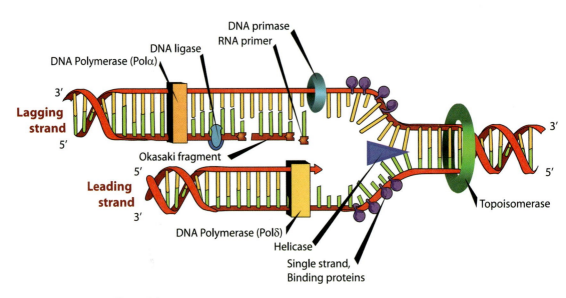

Figure 5.9
A cartoon illustrating DNA replication. A topoisomerase unwinds a bit of double-stranded DNA, and a helicase splits the two strands apart. A set of single-strand binding proteins are attracted to the region and act to stabilize the split strands. A primase starts the creation of a complementary strand by putting down an RNA template. On the leading strand, a DNA polymerase III (the delta form) adds complementary bases continuously, forming a pair of complementary strands from the original single strand. The other original strand, called the lagging strand, is handled differently. Since the DNA polymerase only adds nucleotides to the 3′ end of a strand, it has to work backwards. Small segments called Okasaki fragments are synthesized by the repeated action of DNA primase and the alpha form of the polymerase. These fragments are then stitched together by a ligase.

helicase to split the two complementary strands of DNA apart. The place where that split is opened is called a **replication fork**, where copying starts. The replication fork is rapidly surrounded by more proteins that bend the two DNA strands, pulling them further apart. An enzyme called **DNA primase**, inserts a small structure (actually made of RNA) that is the anchor from which the creation of a new DNA strand begins. This primer is extended on its 3′ end by attaching a nucleotide complementary to the next one in the original DNA, through the action of an enzyme called **DNA polymerase III** (which is a **DNA-directed DNA polymerase**). DNA polymerase III continues to add complementary nucleotides to the original strand with a free 3′ end (called the **leading strand**), recreating the original double-stranded DNA. Part of the energy that drives this process comes from the triphosphate bond at the 5′ end of the nucleotide being attached, liberated when the new nucleotide is bonded to the 3′ end of the growing strand. This process continues until the end of the

DNA is reached, or, in the case of circular chromosomes, until the origin of replication is encountered, creating a complete copy of the original DNA.

The other, complementary strand of the original DNA also needs to be made double-stranded again, through the addition of complementary nucleotides. However, since the polymerase operates by adding nucleotides to the 3′ end of a strand, this process has to work "backwards." As this, the **lagging strand**, is pulled apart from the original DNA, DNA primase lays down primers, which are then extended a little ways back toward the replication fork, making fragments of replicated, double-stranded DNA (called **Okazaki fragments**) interspersed with RNA primers. Eventually another enzyme, called an **exonuclease**, removes all the RNA primers. Finally, DNA polymerase fills in the missing nucleotides and yet another enzyme, called **DNA ligase**, seals those nucleotides to the rest of the backbone. When the leading and lagging strands have both been made double stranded, two identical, double-stranded DNA molecules result; the genome has been duplicated.

This rather complicated and baroque process is universal throughout cellular life,[11] and can happen very quickly. In bacteria, it can proceed at a rate of more than 1,000 duplicated nucleotides per second. It is also highly accurate. In roughly 1 out of 10,000 replications, an incorrect nucleotide binds, potentially causing an error. However, DNA polymerase III has a "proofreading" capability, which catalyzes the removal of incorrectly base-paired nucleotides at the 3′ end of a strand, driving the final error rate to roughly 1 in 1,000,000,000 nucleotides. Other mechanisms for repairing DNA damage are discussed later.

A mutation rate of one in a billion might seem pretty negligible, but since even small genomes contain millions of nucleotides, each replication has a greater than one in a thousand chance of containing an error. An error in the replication of a nucleotide in a DNA molecule (which results in a **single nucleotide polymorphism**, or SNP, pronounced "snip") is an important category of mutation, and creates the variation in alleles described functionally in section 2.2. In a large population and over many generations, a large number of SNPs will occur.

A SNP can have various consequences for the offspring. Since most amino acids are specified by more than one codon, some SNPs don't change the resulting gene product at all, making them **silent mutations**. SNPs that occur outside of protein coding or regulatory regions likely have no effect on phenotype. Such mutations are neutral, in that they have no effect on the fitness of an organism.

---

11. Some viruses store their genomes as RNA, but their replication depends on hijacking a cell's nucleic acid processing machinery, and all cells use this DNA process.

Neutral SNPs are important for the understanding of biology, however, since they accumulate at random and are inherited by all descendents. Tracking the patterns of neutral mutations in groups of organisms often allows for the reconstruction of the evolutionary history of a population. This sort of analysis has transformed the understanding of the evolutionary history of species as well.

Some SNPs do cause an amino acid substitution in a protein. Changing a single amino acid in a protein sometimes also has no discernable effect, and might be neutral. For example, pig insulin works well in humans, despite a SNP that results in the change of an amino acid; it was used to treat diabetics until fairly recently.

The variations that drive evolution, however, are the ones that are not neutral. A change in the amino acid sequence of a protein sometimes does affect the functioning of the molecule. Most often, the change of function is negative (deleterious), the offspring with the mutation will reproduce less successfully as a result, and the mutant allele will be lost from the population over time. Rarely, a mutation will be advantageous, and the increased fitness of the descendents who have the mutation will ensure that the allele is spread throughout the population.

Not all changes in DNA affect a region that specifies a protein. Sometimes a SNP will change **intergenic** (between genes) regions; these mutations are usually neutral. However, some mutations outside of a protein-coding region can have significant evolutionary consequences. Regions of DNA near the beginnings of genes are recognized by the transcription factors that control when a gene is expressed. These regions are called **cis regulatory sequences**,[12] or sometimes **transcription factor binding sites** (TFBS). Mutations in these regions can change the situations in which a gene is expressed, rather than change the gene product. Such mutations often have dramatic effect, particularly on the development of multicellular organisms (see section 7.2.1).

It is also important to note that not all mutations are SNPs, although they are by far the most common. Nucleotides or whole regions of DNA can be lost (called **deletions**) or gained (**insertions**). Insertions and deletions can involve large stretches of DNA. One particularly important sort of insertion is **gene duplication**, where an extra copy of a region containing at least one entire gene is erroneously added to a chromosome. Though very rare, they have often turned out to be advantageous.

---

12. **Cis**, pronounced "sis," means "on the same side," and is the opposite of **trans**, or "on the other side." Here, cis indicates the regulatory signal is in the same molecule (DNA) as the thing being regulated.

Sometimes gene duplications are advantageous because a lot of a gene product needs to be created at once. For example, among the most duplicated genes are the ribosomal RNAs. In fact, some frogs have 20,000 more or less identical copies of a ribosomal gene in their genome. How is this advantageous? Cells need to produce a lot of protein, and ribosomes are where that protein is synthesized. It is not unusual for a cell to have 10 million ribosomes, and that many new ones need to be made every time a cell divides. The rate at which ribosomal genes can be transcribed could be the rate-limiting step in protein synthesis, so organisms that are able to transcribe certain genes, like ribosomal RNAs, in parallel could have an advantage.

Most gene duplications that persist do not involve such large numbers, but there are other reasons they can be advantageous. Duplications usually begin as neutral mutations, since having an extra copy of a gene has little effect. However, the functions of duplicated genes can diverge from each other over time; one copy can accumulate variants without harming the functioning of the other, ultimately resulting in two or more specialized versions of the original gene. For example, humans have seven different hemoglobin genes; two are used only during fetal development. Genes within a single organism that have a common ancestor are said to be **paralogs**, and together they form a **gene family**. In most multicellular organisms more than half the genes in the organism will have one or more paralogs. Sometimes, duplications result in genes that are never transcribed, perhaps because of disruptions of upstream regulatory binding sites. These nonfunctional paralogs are called **pseudogenes**. Other large-scale changes in the structure of DNA, such as inverting a region or reordering several regions, also occur, although more rarely.

Living things have evolved a remarkable array of **DNA repair mechanisms** that can detect and fix many sorts of problems. Perhaps the simplest repair is when a nucleotide on one strand is damaged (e.g., by oxidation). **Template-directed** repair uses the other, undamaged strand to excise a damaged nucleotide and replace it with a nucleotide complementary to the other strand. More dramatic damage, such as a double-stranded break, can also be repaired. There are enzymes that detect complementary ends of DNA sequences and ligate them together. This requires short overhangs of matching single-stranded DNA at the ends of the break to identify the pieces as belonging together. This sort of repair is imperfect, and can lead to deletions or insertions. Another sort of repair mechanism, called **homologous recombination**, takes advantage of the fact that, in diploid organisms, two chromosomes have the same genes (although possibly different alleles). If one chromosome is damaged badly (say, by a double-stranded break), the other can be used as a sort of template to repair it. In fact, any DNA sequence that is similar (homologous) over a long enough stretch with a sequence in another molecule might be switched (recombined). A very similar mechanism underlies the crossover that switches between parental chromosomes in meiosis (see section 2.1), and is used by genetic engineers to introduce new versions of a gene into an organism by design (see section 10.3).

Other mechanisms can have a large-scale effect on DNA. One particularly interesting phenomenon is "selfish DNA" that just replicates itself, without any apparent benefit for the organism whose DNA it appears in. One important example of selfish DNA is the **transposon**, a sequence of DNA that manages to be good at copying itself into new locations. Transposons generally code for a protein (naturally called transposase) that splices the transposon sequence into a new place in the DNA, sometimes going through an RNA intermediate. Transposons can be remarkably effective at replicating themselves: nearly half of the human genome is transposons or their remnants. The combination of transposons (which make copies of themselves throughout many genomes) and homologous recombination (which depends on sequence identity) can lead to large-scale DNA rearrangements, including **transpositions** (movements of large segments of DNA in a genome) and **inversions** (changing the direction of a segment), as well as insertions and deletions. The large-scale differences among species, such as changes in the number of chromosomes, may have resulted from such rearrangements.

5.3   Macromolecules and Life

The molecules and processes described in this chapter are found in all living creatures. Though there are many variations to be found (e.g., people replicate their DNA at only about 80 nucleotides per second, more than a dozen times slower than bacteria), this universality is a powerful statement about the ancestry of all living things. The last universal common ancestor, living at least 2 billion years ago, must have had DNA genes that coded for metabolic proteins.

Study of biological macromolecules has opened the door to understanding how unexceptional chemistry can result in such remarkable function. Perhaps even more important, biological macromolecules are permeated with evidence of our evolutionary history. Similarities in nucleotide or amino acid sequences from different organisms are evidence of common ancestry, of **molecular homology**. The patterns of SNPs in such homologous genes can pinpoint the ordering and sometimes even the date that different species diverged from common ancestors.

The macromolecular structures and functions described here are complicated (and still just scratch the surface of what is known), but understanding them opens the door to molecular explanations of many otherwise incomprehensible properties of life. And, as we shall see in section 9.3, the molecular understanding of proteins and genes has led to deep insights into the causes of many previously mysterious diseases, and offers new hope for their treatment. Before we can tackle disease, however, it is necessary to explore the structure and functioning of organisms more complex than bacteria.

5.4 Suggested Readings

The material in this chapter is the focus of most textbooks of molecular biology. The three classic textbooks, all excellent, are Bruce Albert's *Molecular Biology of the Cell* (Garland Science), Harvey Lodish's *Molecular Cell Biology* (W.H. Freeman & Co.) and Cooper and Hausman's *The Cell: A Molecular Approach* (Sinauer and Associates). James Watson's *Molecular Biology of the Gene* (Benjamin Cumings) is an edited volume, and lacks the coherence of the others, but is also outstanding. This field changes so quickly that it is always worth purchasing the latest edition of these books. However, these are expensive books that cover much of the same territory, so it's probably worth browsing them at the library before deciding which one would be best for you.

6 Eukaryotes

As described in section 1.2, the most basic split in the tree of life is between bacteria, archaea, and eukaryotes. Eukaryotes include all of the organisms most familiar to people. All plants and animals, and also the fungi (such as mushrooms and yeast), and many microscopic, single-celled organisms generically called **protists** are eukaryotes. Having covered the molecular biology of bacteria in chapter 4 (and having to omit much discussion of the archaea altogether), it is time to turn to the biology of the more familiar eukaryotes.

Like bacteria, eukaryotic cells are enveloped in a membrane, do metabolic work in their cytoplasm, and organize their genome into chromosomes. However, in each of these areas eukaryotes have made significant elaborations.

Under a microscope, the eukaryotic cell looks strikingly different from a bacterium, as can be seen in figure 6.1. Far greater compartmentalization is apparent in them, most notably with the central structure that appears dark in the image, the **nucleus**. Eukaryotes are defined by having cells with nuclei. In addition to nuclei, eukaryotic cells contain a variety of other complex structures, called **organelles**. For example, figure 6.1 also shows many instances of an organelle called the endoplasmic reticulum, stained in red. Each organelle supports a particular function: For example, the nuclei contain a eukaryote's chromosomes, as well as a variety of internal structures that organize and manage those chromosomes. The endoplasmic reticulum packages proteins for transport out of the cell. Many organelles are bounded by their own membranes, which regulate traffic of substances in and out of the organelle. And, although you can't tell from the picture, eukaryotic cells tend to be much larger than bacteria.

The increased structural complexity is mirrored by an increased biochemical complexity as well. Even relatively simple eukaryotes like yeast have more than 6,000 genes, more than twice as many as the bacterium described in chapter 4. Additional metabolic pathways are necessary to produce and manage organelles, to specify transport mechanisms to shuttle molecules among them, and for the

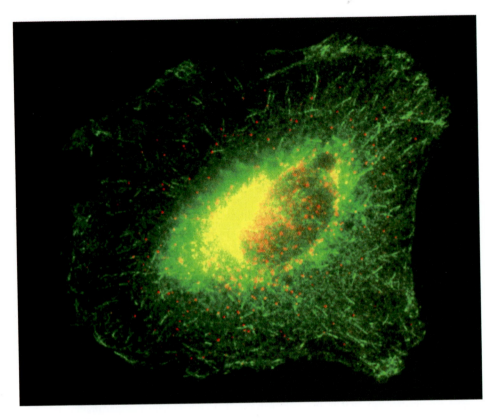

Figure 6.1
A fluorescently labeled human cell. The dark region at the center is the nucleus. The red spots show a portion of the endoplasmic reticulum, an organelle that packages proteins up for secretion; the green tubes are the "highway" of microtubules that transport those packages. The yellow area is a region that is so dense with organelles and microtubules that the colors blend. This photomicrograph was taken by Krysten Palmer, and is used with her permission.

many other functions only eukaryotes employ. The structure of the gene and the control of transcription are also more complex in the eukaryotic cell, which is layered with many additional kinds of molecular processing.

6.1   Eukaryotic Gene Structure and Transcription

A key molecular difference between bacteria and eukaryotes is in the structure of their genes. The coding sequences of eukaryotic genes are interrupted by stretches of DNA sequence that do not code for protein, called **introns**. The mRNA as directly transcribed from the DNA, called **pre-mRNA** in eukaryotes, is processed by a molecular machine called the **spliceosome** to remove these noncoding introns. The spliceosome, itself made of RNA and proteins, recognizes signals in the sequence of the pre-mRNA that indicate the boundaries of the introns. The parts of the gene that are stitched together to form the mature mRNA are called **exons**.[1] In some eukaryotes, particularly humans, several different proteins can be produced from a single gene, through a process called **alternative splicing**. Alternative splices might cause a particular exon to be skipped, or a different exon to be included by the spliceosome, producing a different protein. The exact mechanisms through which alternative splicing is controlled are not yet fully understood. The idea that introns might facilitate the recombination of exons into new proteins is called the exon shuffling hypothesis, and has been suggested as a possible mechanism for generating proteins with novel functions.

The ends of the spliced mRNA are also processed before being transported to the ribosome. Eukaryotic transcripts contain a bit of extra sequence before the start of the coding sequence, called the **5′ untranslated region (5′ UTR)**.[2] The 5′ UTR is capped by a covalent change that protects the beginning of the mRNA sequence. mRNAs that do not have such a cap are chewed up by an enzyme called an **exonuclease**; this is probably a defense mechanism against viruses, which may insert foreign RNA into a cell. The 3′ end of the mRNA, which also has a **3′ UTR**, gets a series of adenine nucleotides, called a **poly-A tail**, added. A 3′ exonuclease slowly degrades RNAs from the 3′ end, so the length of the poly-A tail influences how long the RNA will exist. The entire transcript processing pathway is shown in figure 6.2.

The eukaryotic mechanism for regulating transcription is also more elaborated that in bacteria. Recall that in bacteria, a transcription factor recognizes a signal in the DNA near an operon (a set of genes that are all transcribed together as a unit). In eukaryotes, operons are rare; most genes have their own transcription regulation sites upstream of their coding sequences, called transcription factor binding sites. Each protein having its own transcription factor binding site allows for individualized control over transcription. In addition to the binding sites in the promoter region near the beginning of the gene, eukaryotic genes also have other regulatory sites further away from the coding sequence. These more distant sites are recognized by secondary transcription factors called **enhancers** or **repressors**, which modulate the effect of the transcription factor that binds to the promoter. The expression of a eukaryotic gene is controlled by this complex interplay of combinations of transcription factors. Eukaryotic expression control is illustrated in figure 6.3.

---

1. Noting that **ex**ons are **ex**pressed and **int**rons **int**errupt may help in remembering which is which.

2. Recall from section 4.3 that the two ends of a nucleic acid sequence can be distinguished based on the position of the unbound end in the molecule. The beginning of an mRNA sequence is called 5′(5 prime), and the end is called 3′.

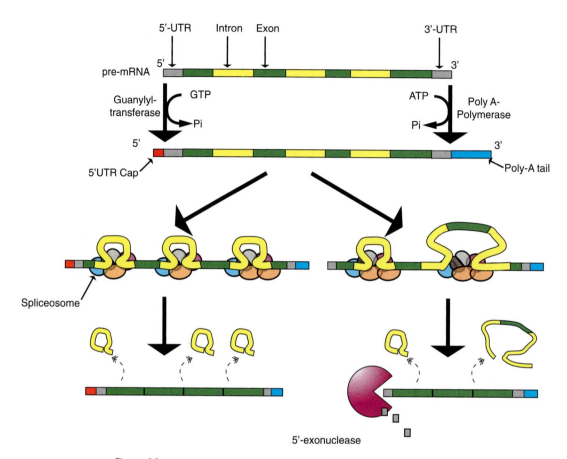

Figure 6.2
The eukaryotic transcript processing pathway. Starting at the top, an immature pre-messenger RNA is shown, containing a 5′ untranslated region, alternating exons and introns, and a 3′ untranslated region. The first step puts a guanine-based cap on the 5′ end of the mRNA and long series of adenines (called a poly-A tail) on the 3′ end; the catalysts and other reactants are indicated in the figure. In the next step, a complex of RNA and protein called the spliceosome causes the introns to be removed. In some proteins and at some times, alternative splicing can lead to different final mRNAs, indicated by the branching arrows. Any mRNA that isn't properly capped will be consumed by a 5′ exonuclease. All mRNAs are eventually degraded by a 3′ exonuclease; the time that takes depends on the length of the poly-A tail.

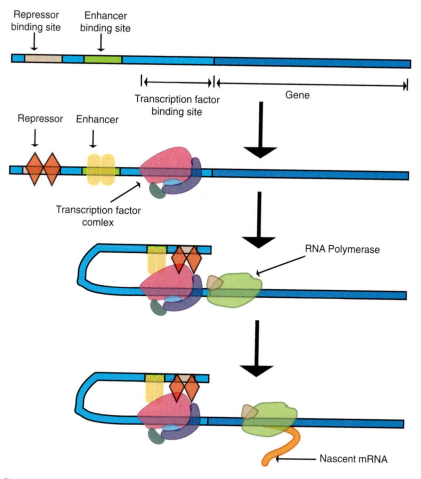

Figure 6.3
Control of transcription of eukaryotic DNA involves many proteins. A transcription factor complex recognizes particular signals, called binding sites, in the DNA upstream of a gene. More distant signals in the upstream DNA are recognized and bound by proteins called enhancers and repressors, which interact with each other and with the transcription factor complex. When all the factors are present in the right combinations, an RNA polymerase is recruited, and begins the transcription of a messenger RNA.

For genes that are to be shut off for long periods of time (e.g., after development) the eukaryotic cell has additional regulatory methods. A covalent modification to DNA, **methylation**, physically impedes access of transcription factors and shuts off the transcription of particular genes. Furthermore, all of the DNA in eukaryotes is tightly coiled around proteins called **histones**, forming a structure called **chromatin**. Unwinding the DNA in one region requires winding more tightly (called supercoiling) in another. The amount of twist in the DNA is actively regulated by enzymes called **topoisomerases**, which also influence transcription. Explicit control of coiling influences whether regions of DNA are physically exposed to transcription factors, as illustrated in figure 6.4. DNA that is accessible to the transcription machinery is called **euchromatin**; DNA that is packed away tightly and inaccessible is called **heterochromatin**. Heterochromatin is also found in chromosomal regions that contain no genes, such as the ends of chromosomes, called **telomeres**, and the centers of the chromosomes, called the **centromeres**.

6.2   Components of the Eukaryotic Cell

The eukaryotic cell has a more elaborated structure than bacterial cells. The cytoplasm is densely permeated by its **cytoskeleton**, a set of proteins that assemble into characteristic filaments that ramify throughout the cell. The most important of these proteins, **actin**, is shown in green in figure 6.1. The cytoskeleton is dynamic and controls the shape of the cell. In combination with another protein called myosin, actin can cause cells to change shape and to move. The cytoskeleton also forms a sort of highway system along which proteins and other cellular components are transported, and anchors all of the organelles in place.

6.2.1   The Nucleus

The activity of transcription regulation and mRNA processing takes place in the nucleus. The nucleus of a eukaryotic cell is its largest organelle, occupying about 10% of the volume of the cell. The nucleus is physically defined by a pair of membranes that separate its contents from the rest of the cell. The outside of the outer membrane is covered in ribosomes. Like the cell membrane, the nuclear membrane contains pores that are selectively permeable, allowing the free flow of water and ions, but restricting the flow of macromolecules. Active transport mechanisms direct the import of transcription factors and export of mature mRNAs, among other things ensuring that immature mRNAs do not reach the cytoplasm.

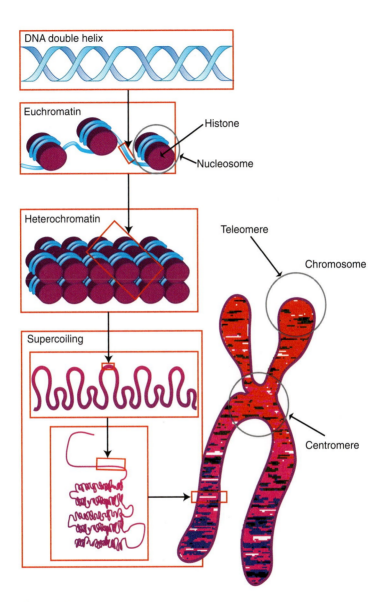

Figure 6.4
DNA packing in eukaryotes. All of the DNA in eukaryotes is wrapped around protein assemblies called histones, forming nucleosomes. These nucleosomes can be accessible to the transcription machinery; if so, they are called euchromatin. More tightly packed and inaccessible DNA is called heterochromatin. These regions of DNA are further twisted together to form supercoiled regions, and ultimately packed up into chromosomes. The tips of the chromosomes are called telomeres, and the middle is the centromere.

Nuclei themselves have some internal structure. Clearly visible under the microscope as a dark spot within the nucleus is the **nucleolus** (nu-cle-oh-lus), which is a specialized structure for creating ribosomal RNA and assembling ribosomes. The nucleolus mostly consists of the genes for making rRNAs.

6.2.2 The Endoplasmic Reticulum and Golgi Apparatus

Contacting the outer nuclear membrane is an organelle called the **endoplasmic reticulum**, or ER. The ER is a network of tubes and sacks that manages the production and transportation of much of the material within a eukaryotic cell. When the ER is covered in ribosomes, it's called rough endoplasmic reticulum because of its bumpy appearance. Rough ER is dedicated to the production of proteins that are targeted to particular compartments in the cell. The sequences of these proteins include a short region at the beginning of the chain, called a **signal sequence**, that specifies where the protein belongs. After synthesis, the ER packages proteins into **vesicles**, which are transported to the proper location. When properly localized, the signal sequence is removed from the protein by another enzyme. Rough ER also produces proteins that are destined to be excreted by the cell, sending those to an organelle called the **golgi apparatus**, which packages them for excretion (and sometimes for other destinations within a cell). Smooth ER is a related organelle in which the synthesis of lipids and the metabolism of carbohydrates occur.

6.2.3 The Mitochondrion

The mitochondrion (plural is mitochondria) is a particularly interesting organelle. Functionally, it is where most of the cell's ATP is produced, using a process that is much more efficient than the glycolysis described in section 4.2. Structurally, these membrane-bound organelles are notable for having their own DNA. While mitochondria cannot live on their own outside of a cell, they show clear evidence of having once been free-living organisms, related to contemporary proteobacteria. Their DNA is circular, like a bacterium's, and their membranes are much more like bacterial membranes than like those of eukaryotes. The currently best supported theory of the origin of these organelles is through **endosymbiosis**, where two organisms live together in a mutually beneficial relationship, one inside the other, with the eventual loss of some of the mitochondrial functions that allowed them to live independently.[3] Not being coded for by the DNA in the nucleus means that mitochondria have to reproduce independently, which they do. In fact, all of the mitochondria in most animals arise from those that were present in the egg cell, none coming from sperm.[4]

Mitochondria are also interesting for the sophisticated and efficient metabolic pathway by which they produce ATP, variously called the **TCA (tricarboxylic acid) cycle**, the **citric acid cycle**, or the **Krebs cycle**. This is the best studied, and arguably most central, metabolic pathway in all living things. Its use is not restricted to eukaryotes; aerobic (oxygen using) bacteria also use it, so it must have originated before the last common ancestor of bacteria and eukaryotes. The pathway requires the use of oxygen, so it must have arisen after the oxygen-driven mass extinction event of about 2 billion years ago (see section 2.4), although parts of the cycle are also found in anaerobic (oxygen-avoiding) bacteria, and are therefore more ancient even than that. One theory, described later, has a predecessor of the TCA cycle present at the origin of life. However, before considering the implications for the origin of life, a few of the details of the molecular functioning of the cycle are necessary. A lot of new terms and molecules are introduced here, so use figure 6.5 to follow along, and take it slow.

Processing in the mitochondria often begins with pyruvate, the three-carbon compound that is the output of glycolysis (as described in section 4.2). The first processing step is the removal of a carboxyl group from pyruvate through the action of three enzymes, producing a compound called **acetyl-Coenzyme A** (often written acetyl-CoA) and one molecule each of the charged redox intermediates NADH and $FADH_2$. Although a cycle has no beginning and no end, conventionally the beginning of TCA is taken to be where the acetyl-CoA is processed, transferring its acetyl group to the compound **oxaloacetate** thereby forming **citrate**, as shown at the top of figure 6.5. The series of reactions shown in the figure produces five molecules of redox intermediates (NADH and $FADH_2$) that are fed into the mitochondrion's **oxidative phosphorylation** pathway to produce ATP. Going around the cycle, the final reaction winds up producing oxaloacetate, ready to start the process again. The oxidative phosphorylation process uses oxygen as an electron acceptor and

3. Another plausible theory involves bacterial predation. See the discussion of *Daptobacter* in section 6.3.

4. Although sperm need mitochondria to live, all mitochondria in a sperm are destroyed on fusion with an egg. This means it is possible to use mitochondrial DNA to trace maternal lineages back many generations, such as in the work studying "mitochondial Eve," the most recent common ancestor of all human women. See, for example, Richard Dawkins' *River out of Eden*.

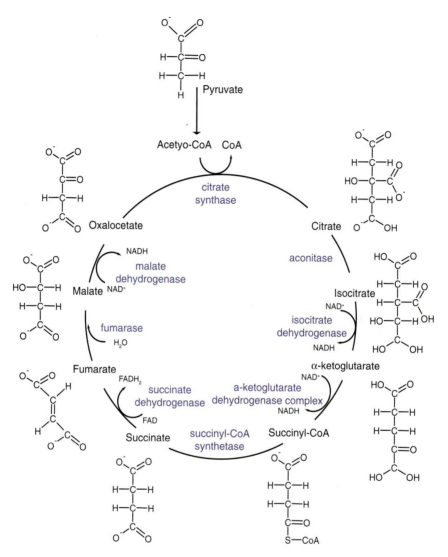

Figure 6.5

The tricarboxylic acid cycle (TCA). This is the central metabolic pathway in all oxygen-consuming living things. The names and chemical structures of each substrate are shown. The dark lines indicate the reactions, the names inside the circle identify the protein catalysts for each reaction, and the curved arrows show other reaction products. See the text for details.

creates a pH gradient (more positive H^+ ions on one side than the other) across the mitochondrial membrane. That membrane potential provides the power needed by the ATP synthase enzyme to produce ATP.

Note that glycolysis creates two molecules of pyruvate per molecule of glucose, and that the conversion of pyruvate to acetyl-CoA produces an additional NADH and $FADH_2$, so in total this process produces 10 NADH and the 2 $FADH_2$ per molecule of glucose. Oxidative phosphorylation can turn these 12 molecules into a maximum of 26 new ATPs (sometimes a bit less due to leakage of H^+ across the membrane). Add in the net gain of 2 ATPs from glycolysis, and counting the 2 GTPs produced by TCA (which can be used directly or converted into ATP), and the entire process can produce a maximum of 30 ATPs from one molecule of glucose, a huge increase in efficiency over glycolysis alone. The waste product is 6 molecules of CO_2, one from each of the 6 carbon atoms in glucose. The use of oxygen and the production of CO_2 underlies the name of this entire process, **cellular respiration**.

Many sorts of energy-rich compounds can be fed into the TCA cycle, not just pyruvate. Fats (triglycerides, see section 3.3) can be used once they are hydrolyzed, which breaks up the fatty acids and the linking glycerol. Glycerol is converted into glucose by a pathway called **gluconeogenesis**, and fed into the glycolysis pathway. The fatty acids are transported to mitochondria and broken down in another energy-generating process called **beta oxidation**, which also produces acetyl-CoA that can be fed into the cycle. Even amino acids can be fed into the cycle to produce energy from proteins if need be; however it is more common in nonstarvation conditions for the cycle to be used to produce rather than consume amino acids, as described later.

The centrality of this process for nearly all living things suggests that it arose very early in the history of life. One particularly compelling account is captured in the phrase, attributed to Harold Morowitz, that "metabolism recapitulates biogenesis." That is, the origin of life is reflected in contemporary metabolism. Of course, very little oxygen was available in the early atmosphere (see section 2.4), and early life forms are hypothesized to have run the TCA cycle backwards: CO_2 and water would be used to synthesize a wide variety of carbon compounds, with hydrogen or sulfates as the electron donors. There are a few contemporary bacteria that live that way (the green sulfur bacteria) and some recent work has shown the potential of this idea in prebiotic conditions.[5]

As important as the TCA cycle is for producing ATP, it also has many other roles in the cell. All of the pathways that bring nitrogen into an organic

5. See suggested readings by Harold Morowitz and Scott Martin for more details.

molecule, and are therefore crucial for the synthesis of amino acids and nucleotides, involve a TCA intermediate. The process of attaching a nitrogen-containing chemical group (generally taken from ammonia) is called **transamination**. The primary substrate for transamination is **alpha-ketoglutarate**, which produces the amino acid glutamate. The other substrates for bringing nitrogen into the system are oxaloacetate (producing aspartate) and pyruvate (producing alanine). All of the other amino acids and nucleotides are synthesized by further processing of these initial compounds.

Mitochondria also have other functions in the cell, for example, regulating cellular redox potential, and, in some cells, detoxifying excess ammonia concentrations. Also, as described in more detail in section 9.1.3, disruption of the mitochondrial membrane triggers an active process that kills the cell.

6.2.4 The Chloroplast

A substantial proportion of the eukaryotes in the world, the plants and green algae, get their energy directly from sunlight. Interestingly, these eukaryotes have another organelle that contains its own DNA and directs the production of energy, called a **chloroplast**. Chloroplasts also likely had an endosymbiotic origin, as their DNA shows homology with free-living photosynthetic bacteria, cyanobacteria.

Inside the chloroplast are stacks of disklike structures called **thylakoids**, surrounded by their own membrane. This is where photosynthesis, transforming light into ATP, takes place. The biochemical mechanism by which light is transformed into energy, called the **Z-scheme**, is complex, involving hundreds of proteins, and is illustrated schematically in figure 6.6. The basic outline involves three molecular machines that form a multistage electron transport chain centered on the pigment **chlorophyll**. The first machine, called the **antenna complex**, is a set of proteins and light-absorbing pigments that can capture light at several wavelengths, transform it into the narrow wavelength that chlorophyll is sensitive to, and couple it to a chlorophyll molecule in the next molecular machine, the type II **photosystem**. When a reaction center in that chlorophyll is illuminated, two electrons are excited to a higher energy level and are then captured by the electron acceptor protein **pheophytin**. To recharge the pigment, two electrons are captured from water, either by photolysis or through enzymes, releasing oxygen as a byproduct. This ability to split water into hydrogen and oxygen is one of the key chemical innovations of photosynthesis.

An electron transport chain transfers the activated electrons through the **cytochrome b6f complex** of four proteins. The cytochrome b6f complex uses some of the energy to pump H^+ ions across the membrane enclosing the

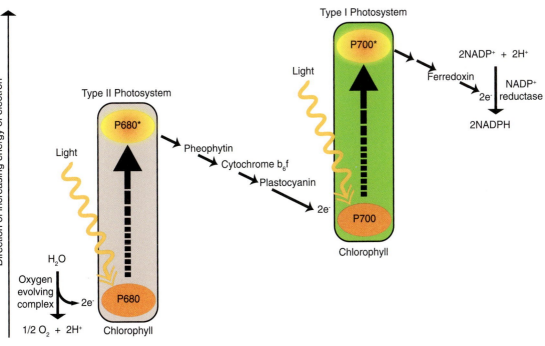

Figure 6.6
The Z-scheme, the biochemical mechanism through which plants transform light into energy. Starting at the left of the diagram, light channeled to chlorophyll captures two electrons from water (splitting it into hydrogen and oxygen) and excites them to a higher energy level in the type II photosystem. Those electrons are transported through a chain starting with pheophytin, capturing some of that energy, which is then used to produce ATP. The electrons then charge another chlorophyll molecule in the type I photosystem, passing through another transport chain, producing NADPH.

thylakoids. The osmotic flow of these ions back through the membrane drives ATP synthase to produce ATP, the same way it does in mitochondria. The ultimate destination of these electrons is to recharge another chlorophyll molecule in the type I photosystem. When the type I photosystem chlorophyll is illuminated, the activated electrons follow a different transport chain, through the protein ferredoxin, ultimately providing the energy needed for an enzyme called NADP reductase to produce NADPH.

The NADPH and ATP created in photosynthesis are used to power the **Calvin cycle**, the pathway by which plants are able to create all of the molecules they need from CO_2, shown in figure 6.7. Briefly, the cycle depends on the enzyme **RuBisCO**, which integrates environmental CO_2 with a five-carbon compound called **ribulose bisphosphate**, which is then split into two molecules of **3-phosphoglyceric acid**, or 3PGA. Two additional steps, consuming

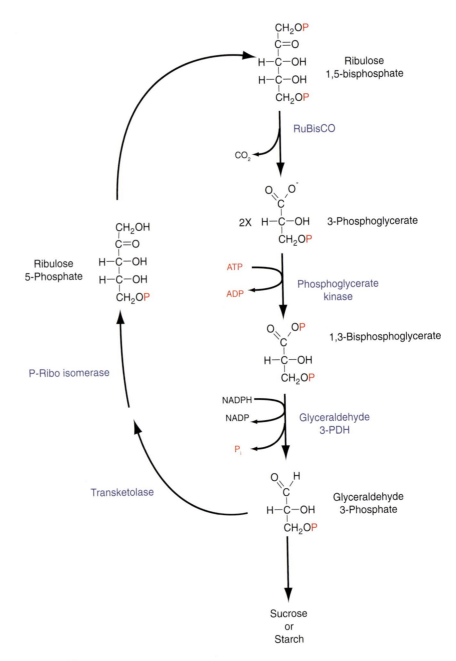

Figure 6.7
The Calvin cycle, the pathway by which plants create biomolecules from carbon dioxide. The main step is the integration of carbon dioxide into ribulose bisphosphate by the action of the enzyme RuBisCO, which is used to create molecules of glyceraldehyde 3-phosphate or G3P. G3P is used to produce both the sugars and starches the plants need, and to create more ribulose bisphosphate, to keep the process going.

an ATP and an NADPH, transform 3PGA into **glyceraldehyde 3-phosphate**, or G3P. In photosynthesis, G3P occurs on the path to producing sugars and other materials, primarily starches and polysaccharides, that the cell needs to make stems, leaves, and other structures. (Recall from section 4.2 that in sugar-consuming organisms G3P is an intermediate compound in glycolysis, and is broken down to produce energy.) Through a different set of additional reactions G3P is also the substrate used to produce more ribulose bisphosphate, closing the cycle.

6.2.5   Other Plastids

Chloroplasts are a kind of plastid, which is the generic name for plant organelles that have their own DNA and are likely to have arisen endosymbiotically. Chromoplasts are the organelles that produce and store pigments, including the ones that give flowers and fruits their colors. Leucoplasts produce and store **terpenes**, a class of specialized molecules made only by plants, including the **essential oils** that underlie most natural (and now artificial) fragrances. Specializations of the leucoplasts include organelles for producing and storing particular oils, starches, and proteins used by plants, as well as the **statolith**, an organelle that a plant uses to detect the direction of gravity.

6.2.6   The Proteasome, the Lysosome, and the Peroxisome

Proteasomes, **lysosomes**, and **peroxisomes** are different cellular components that are responsible for the destruction of damaged or unnecessary molecules. The proteasome is a widely dispersed molecular machine that degrades proteins. It is crucial that the proteasome chews up only proteins that are genuinely damaged or unnecessary. A variety of mechanisms can mark proteins for destruction by attaching a small protein called **ubiquitin** to them. Once a single ubiquitin has been linked to a protein, it attracts others, and the protein becomes **polyubiquitinated**. A protein with four or more ubiquitins will be transported to the proteasome and chopped up into small pieces. Ubiquitinization is used to regulate protein concentrations as well as to prevent damaged or misfolded proteins from doing harm.

Lysosomes and peroxisomes are membrane-enclosed organelles that also degrade other molecules. The insides of both organelles are packed with digestive enzymes and are highly acidic (i.e., they have low pH). They are able to break down proteins, nucleic acids, and polysaccharides. Lysosomes can also digest bacteria that manage to invade a cell. Peroxisomes specifically target compounds that pose an oxidative hazard, particularly hydrogen peroxide (H_2O_2). In addition, the peroxisome is involved in the process of beta oxidation (described in section 6.2.3), which is a step in the extraction of energy from fats.

6.3 Lifestyles of the Single-Celled Eukaryote

Detailed consideration of the familiar eukaryotes, plants and animals, must await the discussion of multicellular organisms in the next chapter. However, even single-celled eukaryotes, generically called protists, have a much richer set of capabilities and behaviors than bacteria, starting with the way they reproduce.

All eukaryotic cells go through a repeating sequence of events called **cell cycle**. The cell cycle is divided into distinct periods, called phases. The most basic division is between **M phase** (named for mitosis), when the cell is physically dividing, and **interphase**, which is the rest of the cycle. Interphase is further divided into three phases: **G1** (gap 1) **phase**, when the newly formed cell undergoes rapid growth, synthesizing the compounds necessary for division; **S** (synthesis) **phase**, when the chromosomes are duplicated in preparation for division; and **G2** (gap 2) **phase**, when final preparations for division are made. At the transitions between each of these phases are **checkpoints**, where the integrity of the DNA and other aspects of the dividing cell are tested. Failure at a checkpoint causes the cell to go into **arrest**, and cease reproducing. Some cells in multicellular organisms (e.g., neurons) eventually cease dividing altogether, but continue in a growth phase called G0.

M phase is also subdivided into four phases: **prophase**, when the duplicated chromosomes condense, are physically separated from each other, and recombination occurs; **metaphase**, when the separated and recombined chromosomes are aligned perpendicularly to a structure called the **spindle**; **anaphase**, when protein motors pull the chromosomes away from the spindle toward opposite ends of the nucleus; and finally **telophase**, when the newly separated sets of chromosomes are repackaged in separate nuclear membranes and the chromosomes decondense, preparing again for interphase.

Even single-celled eukaryotes are sometimes capable not only of mitotic division, but also of the sexual recombination of meiosis. Consider the case of the brewer's yeast, **Saccharomyces cerevisiae**, one of the primary experimental organisms in which the molecular biology of eukaryotes has been studied. When subject to stresses, such as nutrient deprivation, these normally diploid yeasts will divide into four haploid cells of two types, labeled a and α (male and female perhaps seems too strong for protists).[6] The haploid form can reproduce by mitosis, creating more haploid yeasts; the descendents can even switch between the a and α mating types while remaining haploid. However, the haploid yeasts have elaborate mechanisms that drive them back toward diploidy. The haploids excrete a protein, called a **pheromone**, specific to their mating type. Yeasts of the opposite mating type have receptors on their cell surfaces that recognize this signal, causing the cells to grow together, fuse, and become diploid again. These signaling molecules and receptors that underlie the yeast's search for sex partners have since evolved to become a critical aspect of the development and function of multicellular organisms, and still play many roles in the human search for mates.[7]

Other protists have also been shown to switch to sexual reproduction in times of stress. How sexual recombination might have arisen in the history of evolution, and why it is selectively advantageous, are still open questions in biology. Sex clearly has a cost, as sexual reproduction reduces by twofold the number of offspring,[8] so any advantage accrued has to be immediately significant. There are two broad classes of hypotheses regarding the origin and advantages of sexual reproduction: one focuses on the value of diverse offspring, and the other on the value of another genome for repairing damaged DNA. Perhaps the most surprising of the diverse offspring theories is the suggestion that sex evolved to confer resistance to parasites.[9]

In addition to sexual reproduction, eukaryotic protists also gained more elaborate abilities to sense the environment, move, and accomplish other sorts of actions. One of the most evolutionarily significant of these innovations was predation. No fossils remain from the first microorganism that subsisted on others, although there are a few modern-day bacteria that are clearly predatory.

---

6. Recall from section 2.2 that diploid cells have two copies of each chromosome, and haploid cells have only one copy. In many organisms, the reproductive cells (e.g., sperm and eggs) are haploid, and all other cells are diploid. In protists, the haploid state generally has the potential for sexual combination with another, whereas the diploid state does not.

7. For example, the proteins called G coupled protein receptors (GPCRs), are homologs of yeast pheromone receptors. GPCRs are the central molecules that endow humans with the ability to see and smell. Their role in development is described in section 7.2, their role in sensation is described in section 8.3.3, and their importance to the modern pharmaceutical industry is described in section 9.7.

8. In sexual reproduction, the unit of reproduction is the couple; in asexual reproduction it is an individual. Unless a couple can produce more than twice as many surviving offspring as an asexual individual, sexual reproduction will have a lower reproductive success.

9. This is the so-called Red Queen hypothesis. If parasites have short generation times compared to their hosts, they will likely adapt to host responses more quickly than the host can produce them. Sexual recombination makes much larger changes in hosts, altering that coevolutionary balance. See, for example, Matt Ridley's *The Red Queen: Sex and the evolution of human nature*.

One, called daptobacter, uses a flagellum to gain speed (it's the fastest known bacterium), and then collides with, penetrates, and grows inside of its prey, suggesting a mechanism that may have played a role in the origin of eukaryotic organelles.

Predators and prey are particularly salient aspects of the environment to each other, and the threat of being eaten (or starving) generates extreme selective pressure. Predator-prey coevolution can lead to "arms races" of alternating innovations in attacks and defenses. Among protists there are a very large number and variety of predators, and a corresponding variety of defense mechanisms.

The paramecium shown in figure 6.8 conveys the increasingly specialized organization that animal-like protists bring to the single-celled world, showing adaptations for both predation and defense. Paramecia are covered with small hairs, called cilia, which are attached to chemical motors just inside the cell membrane; this mechanism allows the cell to move with some directional control. Specialized cilia sweep bacteria into an opening in the cell called the

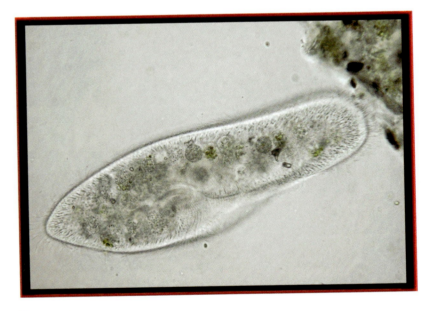

Figure 6.8
A paramecium, a single-celled, animal-like protist. The opening on the bottom toward the left is the gullet into which the hairlike cilia sweep the bacteria that are the paramecium's food. The dark structures just inside the cell membrane are trichocysts, a defensive adaptation that shoot out when the paramecium is threatened, making it difficult to consume. This photograph was taken by Ron Neumeyer from his beautiful Web site at http://microimaging.ca, and is used with his permission.

gullet. When enough bacteria have accumulated in the gullet, it pinches off and becomes a food **vacuole** (vacuole is a generic name for a membrane-surrounded internal structure). Digestive enzymes are introduced into the vacuoles, and nutrients are exported into the cytoplasm. Eventually, when the vacuole contains only remnants that cannot be further exploited, it fuses with a specialized structure in the cell membrane called the anal pore and the waste is ejected. Paramecia themselves are prey to other creatures, and so they have developed a remarkable defensive system based on structures called trichocysts, visible just inside the cell membrane. When a paramecium is attacked by one of its predators, thousands of trichocysts shoot out very rapidly, making it difficult for a predator to attach to the cell.

Single-celled eukaryotes are remarkably complex organisms, and their molecular biology is in many ways very similar to our own. However, all of the creatures that are familiar to most people, even the smallest ones, are composed of more than one cell. Most of the biological topics directly relevant to people, from childhood development to the diseases of old age, hinge on understanding the structure and function of multicellular organisms.

6.4   Suggested Readings

Driven by new molecular techniques, scientific appreciation and understanding of the world of single-celled eukaryotes is growing rapidly. *Genomics and Evolution of Microbial Eukaryotes*, edited by Laura Katz and Debashish Bhattacharya (Oxford University Press, 2006), gives an excellent overview of these recent advances.

There is a large literature devoted to theories of the origin of life. Robert Hazen's *Genesis: The Scientific Quest for Life's Origins* (Joseph Henry Press, 2005) is an even-handed overview of many approaches to the topic. Harold Morowitz's book *Beginnings of Cellular Life: Metabolism Recapitulates Biogenesis* (Yale University Press, 2004) describes just one of these theories, but it is clearly written, brief, and quite compelling. Scot Martin's work, described on http://www.seas.harvard.edu/environmental-chemistry/projects/prebiotic.php provides empirical support for Morowitz's theory.

The origin and evolutionary advantages of sexual reproduction remains a fascinating and open topic. The scientific publisher *Nature* devotes a Web site to this question at http://www.nature.com/nrg/focus/evolsex and published a special issue of the journal *Nature Reviews Genetics* in April 2002 (vol. 3, No. 4) dedicated to the evolution of sex. Matt Ridley's classic book, *The Red Queen: Sex and the Evolution of Human Nature* (Harper Perennial, 2003, originally published in 1993) remains an excellent and accessible introduction to the field.

It may seem unlikely for a popular book to be dedicated to an organelle, but the mitochondria are central to many human concerns, including aging and gender differences. Nick Lane's *Power, Sex, Suicide: Mitochondria and the Meaning of Life* (Oxford University Press, 2005) is a remarkable effort, making clear that the function and history of these components of all eukaryotic cells account for a great deal of why we are who we are.

7 Multicellular Organisms and Development

Armed with a sense of the biochemistry and cellular architecture of the eukaryote, it is now possible to consider the last major transition on the way to familiar plants and animals. These organisms are all **multicellular**, meaning their bodies are made up of anywhere from dozens to tens of trillions of cells.

When bacteria have access to a good food supply, they can reproduce so well that a colony of them can become large enough to be visible to the naked eye. However, a bacterial colony is not a multicellular organism, even if it consists of a large number of cells with identical genotypes. The hallmark of a multicellular creature is a distinction between the cells whose DNA will direct the creation of another organism, called the **germline**, and all the other cells, called the **soma**. Although somatic cells may divide, they are reproductive deadends; they will not produce another organism. The distinction between germline and somatic cells is evolutionarily profound: A mutation in a germline cell will be passed on to all descendents; a mutation in a somatic cell will die with it.

7.1 The Origin of Multicellular Organisms

Somatic cells make the ultimate altruistic sacrifice, giving up their own reproductive future for the benefit of some other cell. What sort of reproductive advantage could make this apparent suicide advantageous?

Somatic cells **differentiate**, becoming specialized for particular tasks. Coordination among specialized cell types makes possible the remarkable capabilities of tissues and organs and bodies, dramatically increasing the reproductive success of the germline cells that gave rise to them. The functional benefits of cellular specialization are illustrated by all of the organisms you can see. Consider a (simplified) example from a land plant: A root cell is specialized for absorbing water and other nutrients from the dirt, and resisting the sorts of dangers found underground, but it can't produce the energy it needs, since it can't photosynthesize. A leaf cell out in the sunlight is dedicated to photosynthesizing and can produce a huge amount of ATP, but it needs a steady supply of water and other nutrients that aren't readily available above ground, the best place to do its photosynthetic work. Specialized cells working together can exploit multiple niches simultaneously, gaining efficiency that gives them a reproductive advantage over the same number of independent cells.

A key question in the history of life is how this grand deal could have gotten started. Each step in any evolutionary path has to be independently advantageous. What sort structure could have arisen in a single-celled ancestral organism that would lead to functions of multicellularity? The issue is illuminated by the "biographies" of two organisms that are just barely multicellular: the slime mold Dictyostelium and the green algae Volvocales.

Dictyostelium, shown in figure 7.1, mostly lives as a protist, feeding on bacteria. However, when food runs short, a unicellular Dictyostelium will start producing a chemical signal that attracts other nearby Dictyostelia. As a result, as many as 100,000 cells join to form an aggregate. This aggregate acts as a unit, sending and receiving chemical signals among each other that characterize the environment and cause the group to move in a coordinated direction. More important, the cells specialize into two categories, some becoming a stalk that acts as a scaffold to raise up the others. The stalk is topped by a fruiting body that releases spores. As much as 20% of the colony will die making the stalk, and forgo the opportunity to reproduce at all. The cells in the stalk have given their lives to increase the reproductive success of the others in the colony.

Although natural selection is, of course, still doing its inexorable work on the individual Dictyostelium cells, an additional kind of selection is also happening at the level of colonies. The cells in colonies that produce a stalk are more reproductively successful than cells that are not in such colonies, enough so to make up for the loss of the stalk cells. Selection is acting on the group, as well as on the individuals. The level of selection has changed.

Group (or sometimes **multilevel**) **selection** is an idea with a controversial history, and many hypotheses that invoked it have turned out to be wrong. However, in a number of cases (including colony formation, multicellularity, and the behaviors of certain kinds of social insects) there is clear evidence for changes in the units of selection.[1]

---

1. For an accessible account of this evidence, see John Maynard Smith and Eörs Szathmáry (1999), *The origins of life: From the birth of life to the origin of language.*

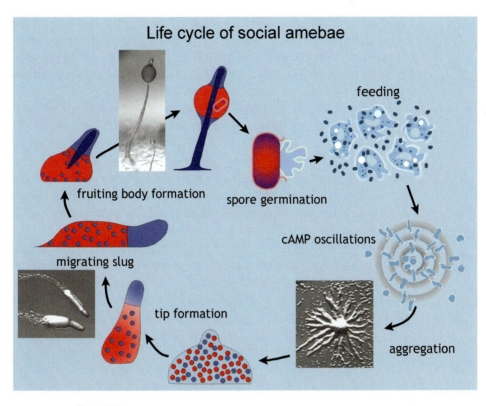

Figure 7.1
The life cycle of a Dictyostelium. Starting at the top right and going clockwise: Dictyostelium generally live as independent protists, feeding on bacteria. When food runs short, they send out pulsed chemical signals that cause other Dictyostelium to aggregate, and then to differentiate into stalk and spore cell types. The stalk cells concentrate, and lead the other cells toward a good place for releasing spores. Those cells form a stalk that lifts the spore cells, which then release reproductive spores into the air. Spores that land in richer environments then start the process anew. This figure was created by Pauline Schaap and is used with her permission.

A key issue that arises when selective advantage accrues to a group is how to avoid **freeloaders**.[2] In the Dictyostelium case, which cells end up being stalk (and not reproducing) is controlled by the concentration of an excreted signaling molecule, and depends largely on the position of the cell in the aggregation. However, a freeloader could fail to respond to the stalk signal, or try to produce its own signals that would cause others to become stalk. If

2. Freeloading is a ubiquitous problem for all living things, and a good path to reproductive fitness for those who can find successful ways to take advantage of other life forms. Parasites and predators can both be seen as kinds of freeloaders. In fact, autotrophs are the only organisms that do not make their living at least partially by exploiting the activities of others.

a freeloader could live in colonies with stalk-forming cells, it could get the benefit of the group activity without ever paying the price. That would make freeloaders more reproductively successful than cells that played by the rules, and the collective activity would fall apart. For a group phenomenon to persist, such conflicts at the lower levels have to be constrained.

One way to constrain conflicts is to ensure that all of the cells in a multicellular group have exactly the same genomes (called "being clonal"). Since it is the reproductive success of the genotype that matters, cells that die in the course of successfully passing along their own genotype (through other cells), are contributing to the success of that genotype. The genotype will be selected for, even though it causes the reproductive failure of some of the organisms that carry it. Alleles that can propagate themselves at the expense of (some of) the organisms that carry them are called **selfish genes**.

In fact, the sharing of the genotype doesn't have to be perfect. **Kin selection** is when related organisms are altruistic in proportion to the degree they are related. An allele that lowers an individual's fitness but increases the fitness of related organisms by a total amount greater than the individual's loss should spread through populations despite its individually deleterious nature. The great evolutionary biologist J.B.S. Haldane once quipped that "I'd lay down my life for two brothers or eight cousins."

Since the cells in multicellular organisms all have the same genotype, kin selection assures us that the sacrifices made by somatic cells are selective so long as the germline cells are able to reproduce more successfully than all of the cells would have been able to acting independently. The freeloader problem doesn't go away, however, since mutations in somatic cells might still lead to freeloaders. In response, multicellular organisms have developed an elaborate and multifaceted set of mechanisms to keep somatic cells from becoming freeloaders (aging is one of those mechanisms, and cancer is one result of their failure; see section 9.6). Before considering defenses against defection, a harder problem has to be addressed: How can one genotype have descendents that are both germline and somatic? What sort of structure could underlie this key differentiation among cells that have the same genotype?

An interesting answer to that question can be found in the biography of the family of green algae called Volvocales.[3] Volvocales are a set of closely related photosynthetic green algae that are found as both unicellular and multicellular organisms. A multicellular member of the family, *Volvox carteri* (shown in

3. Michod, R.E. Evolution of individuality during the transition from unicellular to multicellular life. *Proceedings of the National Academy of Sciences, USA*, 104 (2007), 8613–8618. Available as http://www.pnas.org/cgi/content/full/104/suppl_1/8613.

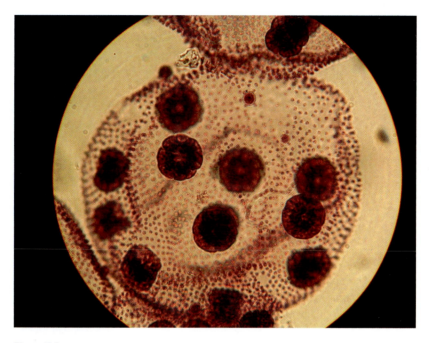

Figure 7.2
A photomicrograph of Volvox carteri. The large dark cells in the center are the reproductive cells; the small lighter cells surrounding them are the somatic cells. This image was created by Aaron Hoffman and Tim Judd, and is used with their permission.

figure 7.2), is about as simple as an organism with germline and somatic cells can be. It consists of about 2,000 nearly identical somatic cells with flagella and up to 16 central reproductive cells without flagella. The selective context in which this division of labor is thought to have occurred is a conflict between the two most basic functions of a living thing: eating and reproducing. While the use of the flagella are required to keep the colony at the optimum water depth for photosynthesis, the flagella themselves are tethered to the same molecular machine that drives cell division. Volvocales can't swim and divide at the same time, so the unicellular ones stop swimming when they divide, falling out of the optimum spot for photosynthesis. The ability to divide the labor between reproductive cells that never swim and swimming cells that never reproduce, linked together as a unit, can provide a meaningful advantage. But how could it have arisen?

The answer can be found in a protein called regA, which is responsible for the differentiation between the two cell types. RegA is a repressor that shuts off the expression of some genes whose proteins play a role in photosynthesis. As a result, the somatic cells have inefficient photosynthesis and don't grow well. Since robust growth is required for mitosis, the lack of growth not only results in the small size of the somatic cells (evident in figure 7.2), but prevents their division. The limited growth doesn't prevent these cells from fulfilling their function, though, which is to use their flagella to keep the organism at the optimal position in the water column.

All of the cells of a particular *Volvox* have the same genotype and arise from a single germ cell. How is it that only some cells become somatic? The germ cell divides mitotically to produce all of the cells in the *Volvox* body. Early in that series of mitoses some of the cell divisions are asymmetric, resulting in large and small cells. The potency of regA depends on the size of the cell. In large cells, the concentration of the fixed amount of regA produced is not high enough to interfere with photosynthesis, but there is a size threshold below which the increased concentration generated by the same amount regA has a major effect on the expression of those genes. The ability to control the size of offspring is well established in many unicellular organisms, some of which conditionally produce small offspring (called **budding**) dependent on environmental circumstances.

Another difficult question is how such an altruistic gene could come into being. What were its precursors, and how where they adaptive? The recA gene can also be found in single-celled Volvocales, and so was present in their common (single-celled) ancestor. How could it be adaptive for an organism to have a gene that would inhibit its own growth and reproduction? The answer is that in a sometimes difficult world, there are going to be environmental circumstances in which temporarily shutting down growth and reproductive activity, holding out for more propitious circumstances, can be adaptive. For single-celled Volvocales, it's not even particularly dramatic: The single-celled algae turn on regA at night, preventing the production of photosynthesis proteins when there is no sunlight. Since producing and maintaining the photosynthetic machinery is metabolically costly, there is an advantage to doing it only when it is beneficial. In many plants and green algae, photosynthesis is adjusted dynamically to respond to environmental conditions. What apparently happened in the case of the *Volvox* is that this regulation was shifted from controlling the *time* when growth and reproduction could happen to control over the *place* in which it could happen, creating two distinguished cell types.

Volvox, like any other group, is susceptible to selfish mutants. Mutations that caused reduced production of or sensitivity to regA would result in the somatic cells regaining their reproductive abilities. However, the original problem hasn't gone away: reproducing cells must lose their flagella, which in somatic cells affects the survival and reproduction of the organism as a whole. Though this selfish mutant appears now and then, it is rapidly outcompeted by the well-functioning multicellular **Volvox** and the organism with the selfish mutant dies out.

The story of the Volvocales illustrates a molecular mechanism that can take a single cell and divide its progeny into multiple paths, even though they all have the same genotype. This control of **cell fate**, and the molecular mechanisms that determine which cells in a multicellular organism will take on what different roles, is at the core of the development of much more complex multicellular organisms.

7.2 Development

All multicellular organisms begin from a single cell, called a **zygote** (usually a fertilized egg, although some multicellular organisms can also reproduce asexually). The process of creating a complete organism from that cell is called **development**. The diverse and fantastic forms of all of the multicellular plants and animals that have ever lived are produced through a remarkably small and well-conserved set of molecular developmental pathways. The broad outlines of the single developmental program that underlies nearly all multicellular life is becoming increasingly clear, thanks both to new molecular techniques that identified homologous development-related proteins across many organisms and to the recent synthesis of embryology and evolution called **evo devo** (**evolutionary development**).

Nearly all multicellular animals (formally called **metazoans**) are bilaterally symmetric; the only exceptions are the Porifera (sponges) and cnidarians (jellyfish, corals, anemones), which diverged from the other animals very long ago. Even some animals that appear radially symmetric (like sea urchins) go through a bilaterally symmetric developmental phase. These rather diverse creatures diverged from each other quite early in the history of life, before the Cambrian explosion, and their last common ancestor was one of the earliest multicellular life forms. Nevertheless, the molecular mechanisms that control the development of all of these animals, called the **Bilateria**, and therefore that ancient common ancestor as well, involve the same few families of proteins.[4]

The molecules of development fall into two broad classes: position-specific transcription factors and cell signaling systems. Transcription factors, proteins that bind DNA to regulate the production of other proteins, are familiar from

---

4. Although the basic ideas are similar in plants, there are some differences. For a sampling of evo-devo approaches to plant development, see Kronk, Bateman, and Hawkinds (eds.), *Developmental genetics and plant evolution*.

the discussions in sections 5.2 and 6.1. Receptor molecules that sense the external world and convey information to the rest of the cell, as well as on the cellular machinery that can release signaling proteins into the environment, were introduced in sections 4.4 and 6.3. These molecular functions underlie the process that forms the animal body, so a bit more detail about how they function is required before describing development.

Cellular activity based on sensing signals in the environment requires three types of functioning. The first function is to recognize particular molecules in the external environment, which is accomplished by receptor proteins that span the cell membrane. Cell membranes in plants and animals are studded with a large number of different sorts of receptors. Each receptor is highly specialized, binding specifically to a particular partner molecule, called its **ligand**, which can be a small molecule or a protein. When the **extracellular** (outside the cell) part of the receptor is bound to its ligand, the **intracellular** (inside the cell) portion of the receptor changes its conformation or some other chemical property.

One example family of receptor molecules is called the **receptor tyrosine kinases** (RTKs). When bound to a ligand, this receptor **dimerizes**, that is, it physically pairs up with another bound receptor of the same type that has also bound to a ligand. The dimerized pair of proteins **autophosphorylate**, meaning that each bound receptor catalyzes the addition of a phosphate group to the other. The phosphate is attached to a particular amino acid, a tyrosine, in each protein, hence the name receptor tyrosine kinase (recall that a kinase is an enzyme that manipulates phosphate groups). Other receptor proteins also use many other mechanisms to can transmit the fact that they have bound to a ligand.

The second signaling function is transmitting the information that a particular receptor has found its ligand to other places in the cell, and integrating the information from multiple receptors, putting together an accurate representation of the external situation. This work is done through **signal transduction pathways** in the cytoplasm, which generally involves addition or removal of one or more phosphate groups on multiple proteins.

In the receptor tyrosine kinase example, the first step in the pathway is made up of sets of **adaptor proteins** that recognize phosphorylated RTKs. One part of the adaptor protein, called an **SH2 domain**, recognizes a particular RTK; another part, the **SH3 domain**, is recognized by a downstream part of the pathway. Through adaptor proteins that mix and match different SH2 and SH3 domains, downstream activity can be controlled by any combination of activated receptors. Such mechanisms can act like logical connectives (AND, OR, NOT), as well as create more quantitative combinations of different receptor inputs.

The duration of response to a signal can also be controlled. The most basic molecular function is a sort of timer switch that changes state when a signal is received and stays that way for some period of time. An important class of these switches is called a **G protein**, the signaling system that yeasts use for finding sex partners, as described in section 6.3. G proteins consist of three different polypeptide chains called alpha, beta, and gamma, and together they bind a phosphorylated nucleotide, **guanosine diphosphate (GDP)** when they are in the "off" state. When a G protein gets the signal from its receptor, the alpha chain releases the other two chains and the GDP. A molecule of **guanosine triphosphate (GTP)** then binds the alpha chain, activating it. Both the released beta-gamma chains and the alpha chain bound to a GTP persist for some time, and activate the next step in the signaling pathway. There are many types of G proteins, with different downstream effects, but each uses very brief molecular events to trigger a state that persists for a period of time, and then resets. The receptors that activate G proteins, naturally enough called the **G protein coupled receptors** (or GPCRs), are among the most important in the body. As described in section 8.3.3, they underlie many human sensory abilities.

Often, the next step in the pathway from a G protein involves the creation of **cyclic adenosine monophosphate**, or **cAMP**. Cyclic AMP, derived from the nucleotide adenosine, acts as a **secondary messenger**, illustrating another important kind of signaling system: one that uses diffusible small molecules as signals. The use of cAMP as a signal arose early; it is the molecule that Dictyostelium cells use to signal to each other to form aggregates. Other small molecule signaling systems are also important. For example, fluxes of ions controlled by membrane channel proteins are also used widely in signaling. The signaling systems used by heart cells and neurons to do their work involve this sort of signaling. All of these mechanisms can interact. For example, cAMP can activate a protein kinase that then can phosphorylate (and therefore activate) its target, and so on.

The final function necessary for a cell to act based on an environmental cue is to transform the processed signal into action, which almost always involves a transcription factor, increasing or decreasing the production of other proteins. In the adult organism, these sorts of signals are the means by which large numbers of cells can coordinate complex activities with each other, and respond appropriately to the situation a creature finds itself in. For example, the rush of "fight or flight" activity triggered by adrenaline is mediated by GPCRs, G proteins, and cAMP, among other players. Also, as described in detail later, cells in the developing organism influence each other's development through the use of such signals. The production, sensation, and integration of intercellular molecular signals are central to both the development and the mature functioning of all multicellular organisms.

7.2.1   Molecular Maps

Only a few hundred genes out of the many thousands in even a small animal genome determine how an embryo will develop. Even more remarkably, these genes are nearly identical in all Bilateria, despite the enormous variations in form. The molecular signals that lay out the form of the human body are the same as the ones that direct the form of, say, a sea urchin or a fruit fly. These highly conserved molecules were among the most surprising findings of the evo-devo work that uncovered the molecular mechanisms of development.

The development of bodies begins with gradients of a few maternal proteins that define the head-to-tail, top-to-bottom, and central-to-distant directions (called **anterior-posterior**, **dorsal-ventral**, and **proximal-distal**). Neither the egg nor the cells of the developing embryo express these genes. They are produced by the mother and anchored at particular spots in the egg cell. Any protein that controls the formation of a body part is called a **morphogen**, and these maternal proteins are powerful morphogens. If an experimentalist changes the gradient of the protein that defines the head end of the cell to radiate from the middle, the embryo will grow a head in the middle of its body.

All known morphogens are transcription factors. Differences in concentrations of those transcription factors in different cells of the developing embryo result in the expression of different proteins in those cells, just as regA does in *Volvox*. Early in the development of the embryo, differential response to various maternal gradient proteins turns on different sets of proteins, called **gap genes**, depending on the position of the cell in the embryo, subdividing the embryo into broad regions. These proteins are themselves transcription factors that interact in complicated ways to form narrow and sharply defined boundaries demarked by all-or-nothing expression of particular morphogens. One reason the interactions are complicated is that they create repeating sequences of morphogens. For example, morphogens encoded by **pair-rule** genes are turned on in precise alternating bands.[5] Repeating sequences of

---

5. These genes and their names are taken from the fruit fly *Drosophila*, in which this system was first characterized. The exact mechanisms of segment formation and the gene names are a bit different for vertebrates, but the general ideas are similar and the molecules are homologs. For more details, see Sean Carroll's *Endless forms most beautiful: The new science of evo devo*. For a more technical discussion, see Alexander Aulehla and Bernard Herrmann's article "Segmentation in vertebrates: Clock and gradient finally joined," in *Genes and Development* (vol. 18 (2004), 2060–2067), available free at http://www.genesdev.org/cgi/content/full/18/17/2060.

morphogens underlay one of the clear commonalities among animals: our bodies are constructed out of sets of repeated parts.

Different organisms have different numbers and kinds of repeated structures, but the modules themselves are widely shared. Consider backbones in vertebrates: Frogs can have fewer than a dozen vertebrae, people have 33, and snakes can have hundreds. Arthropods, the group of animals that includes insects and crustaceans, also have a body divided into different numbers of repeating segments; flies have 14 and millipedes can have nearly 100. The main differences among animals are the size and shapes of these modules, and what is attached to them. The basic modules are all homologs, and the repeated structures within an organism are called **serial homologs**, even when they diverge from each other in exact structure.

The expression of morphogens defines both a set of repeating segments and a sort of coordinate system specifying position in the embryo. The repeating segments generate the basic body pattern, and the coordinate system makes it possible to specify specializations of the segments (e.g., grow legs here). A very important set of coordinate system genes over the anterior-posterior axis is the **homeobox** (**Hox** for short) genes. These genes are highly conserved throughout Bilateria, and are always found, in the same order, in a particular region of a chromosome called the homeobox cluster. Vertebrates have 32 homeobox genes in 4 clusters; most invertebrates have about 8 Hox genes in one cluster. These Hox genes determine what body parts will grow out of which segment. Although Hox expression is linear across the anterior-posterior axis, the relationship is combinatorial: a segment that expresses Hox genes 2 and 3 (they are numbered) will produce a different structure from one that produces just Hox 2.

By manipulating Hox proteins, an experimenter can, say, cause an embryo to grow legs out of its head. More important, through modest shifting of the alignment between the segmented anterior-posterior axis and the Hox genes, the remarkable variety in the number, size, and attachments of segments seen in arthropods and vertebrates can be explained in molecular terms.

Any position in the embryo can be defined as a point in the **morphogenic field** specified by a particular set (and sometimes concentration) of expressed morphogens. Genes that respond to the particular set of transcription factors at a specific point in the morphogenic field cause the cells at that point to begin to go down a path to producing a particular body structure. For example, only at a very specific position in the morphogenic field will a cell begin to transcribe the gene for the pax-6 protein, the master regulator for producing eyes. Once a set of cells starts producing pax-6, the proteins that are then transcribed as a result of its action in turn produce another set of morphogens that create local axes. These local axes define positions within the developing eye, and ultimately invoke the expression of proteins specific to the various parts that go into making up an eye.

One of the surprising findings of evo devo is that the protein that triggers the creation of an eye, pax-6, is highly conserved throughout Bilateria. The amino acid sequence of the protein that causes a developing fruit fly to place an eye in a particular place is quite similar to the one that directs the formation of eyes in, say, clams, snakes, and people. This means that the last common ancestor of all of the Bilateria, living before the Cambrian explosion, must have had an eyelike body part. This commonality among pax-6 is all the more remarkable because the particular parts produced are so different from each other. Fruit flies have compound eyes, made of hundreds of segments with no lens; the nerve cells that sense light are on the surface of the retina in octopi, but on the back side of the retina in people. However, the master regulator morphogen is the same for all: put the mouse version of pax-6 into a developing fruit fly, and it will grow fruit fly eyes, and vice versa. The same is true for the morphogens that trigger the formation of the heart (called tinman), limbs, and many other body parts. The most recent pre-Cambrian common ancestor to the Bilateria must have had a homolog of all of those body parts as well. Other morphogens also control the timing and degree of cell divisions, and the differentiation of cells into particular specialized types. The very ancient origin of the molecular underpinnings of development, and therefore of the basic body plan shared by nearly all animals, is quite striking.

So if the morphogens are all so similar, how is it that the organisms that arise from this developmental process all look so different? Even if, in some molecular sense, lions and lobsters share a body plan, they do have rather different physical bodies. Remember that morphogens are transcription factors, and the transcription control system has both trans and cis parts (see section 5.2.3). The developmental, and therefore morphological, differences between organisms are a result of evolution of the **transcription factor binding sites**, not the transcription factors themselves. Differences in cis regulatory sites shift the expression of a protein in embryological time and place. This is true for the morphogens as well as for other proteins. For example, a change in the transcription factor binding site upstream of a Hox gene shifts the alignment between segments and Hox genes.

This mechanism makes a lot of evolutionary sense. Mutations in the molecular structure of a morphogen would likely interfere with a large number of aspects of development; it would be very rare indeed for all of those changes together to be advantageous, even if one or another of them might have been alone. Changes in the cis regulatory elements can alter the developmental trajectory of one body part without interfering with all the others. Furthermore, a small mutation in the regulatory elements of a morphogen can have a large-scale phenotypic effect, such as described in the discussion of the Sickleback mutation in section 2.2. The mechanism of development based on a spatial map of nearly invariant morphogenic field, with variation arising mainly within the responses to that field has allowed evolution to explore a vast space of possible body plans. Though a small number of gene duplication events in the history of the animals expanded the collection of homeobox and other morphogenic genes used to lay out that map, most of the evolution of animal form has been via changes in regulatory sites, not coding sequences.

7.2.2   Cell-to-Cell Signaling

Not all of the information necessary for development is encoded in a spatial molecular map. The *relationships* among cells in the developing embryo are also important. A simple example can be found in the control of the growth of hair, scales, feathers, and even the compound eye of the fly. These structures need to be placed at regular intervals, but quite close to one another. Although the morphogenic map can be quite finely detailed, these structures are not regulated by some ultraprecise map. Instead, a mechanism called **lateral inhibition** creates these patterns. Each cell that could potentially create the structure excretes a molecular signal that tells the nearby cells not to produce the structure. The effect of receiving the inhibitory signal includes weakening the signal the cell is sending to others, setting up a winner-take-all competition among the cells. Quite rapidly, each neighborhood of cells (defined by how far the signal can travel) settles so that one cell inhibits all its neighbors, creating a regular pattern of roughly equally spaced progenitors. The resulting pattern is created by the local interaction of cells through signaling, not through a global morphogenic map.

The effects of cell-to-cell signaling during development can be quite dramatic. One example is that of a gene family called hedgehog. A product of one of the hedgehog genes is responsible for the lateral inhibition that places hair, feathers, and scales on vertebrates. Another hedgehog gene, called sonic hedgehog, controls differentiation among the developing digits in the vertebrate limb. In people, developmental modules that see high concentrations of sonic hedgehog develop into thumbs and decreasing exposure to the protein (modulated by both concentration and time of exposure) generates the variations among each of the other digits. Hedgehog is a signaling protein, so receptors and a signal transduction system are also associated with it, which ultimately changes the expression of transcription factors based on the cell's molecular environment. There are several other morphogenic signaling systems, and their interactions play an important role in organismal development and body plans.

One of the more remarkable sorts of signals that cells exchange is one that specifies that a cell should die. Programmed cell death, the main form of which is called **apoptosis**,[6] is a critical feature of development. It turns out that many cells in the developing embryo function as a kind of scaffolding. For example, the separation of digits in the vertebrate limb arises because the cells between the digits get an apoptotic signal and die. The development of the small nematode worm *Caenorhabditis elegans* (generally called *C. elegans*) has been studied in great detail, with the fate of every cell from fertilized egg to adult completely elaborated. When the worm hatches from its egg, it has 558 cells; 151 more, nearly 20%, have undergone apoptosis, dying before birth.

Apoptosis also plays a role in the development of systems that are too complex to be completely specified in the genome. There are far more interconnections among neurons than there are nucleotides in the genome, and each of these connections must be wired properly for the nervous system to function properly. In some situations, neurons compete with each other to make functionally important connections, and the cells that don't make the strongest or first effective connection die through apoptosis.

Outside of its developmental context, apoptosis is also used to prevent mutant somatic cells from becoming freeloaders and harming the reproductive success of the organism. The signal to die can come from within the cell itself, from surrounding cells, or even from the immune system of the organism. For example, damage to DNA causes a cell to phosphorylate a protein called p53, activating it. Activated p53 halts the cell cycle (see section 6.3) and causes the production of DNA repair enzymes. If the cell cycle doesn't restart (indicating that the damage has been repaired) after a period of time, p53 acts to initiate apoptosis. This activity prevents damaged cells from reproducing, which is why p53 is called a tumor suppressor gene. Nearly all cancers have defects in the p53 system.

Apoptotic signals are closely tied to cell division signals, and the extent of cell proliferation is governed by the balance between them. The existence and regulation of apoptotic signals is a testament to the fundamentally altruistic nature of somatic cells. The survival of somatic cells does not matter a bit to the forces of selection acting upon them; the only thing that matters is the effect they might have on propagating the germline.

6. Apoptosis is properly pronounced ay-poe-*toe*-sis, since its Greek roots are apo (away from) and ptosis (falling or drooping), although it is commonly heard as ay-*pop*-toe-sis.

Molecular cell signaling plays an important role in adult multicellular organisms as well. **Hormones** are molecules that serve as intercellular signals. Many hormones are proteins (e.g., insulin) or derived from proteins, either by cleaving off a smaller polypeptide signal (e.g., vasopressin, a blood pressure regulator) or by making a posttranslational modification, typically attaching a glucose molecule to the protein (e.g., lutropin, the hormone that triggers ovulation in women). Other hormones are chemical derivatives of amino acids (e.g., adrenaline), or lipids (e.g., testosterone). Hormonal signals are produced and received by every tissue in a multicellular organism.

7.2.3 Cell Motion

The process of creating the body of an animal uses signals that instruct not only what cells do in particular locations, but also their movement within the developing body. In the earliest stages of the formation of the embryo, the body is largely a solid ball of dividing cells, called a **morula**. After a modest number of cell divisions (the exact number depends on the organism), these cells move to form a hollow ball called a **blastula**. These changes in shape are largely driven by **cell adhesion molecules**. Cell adhesion molecules are embedded in the membrane, and determine whether cells will stick to other cells, and, if so, to what kinds. Differences in adhesion combined with random motion can be a powerful generator of specific cellular architectures.

Cadherins are one important class of these adhesion molecules. They are membrane-bound proteins, but unlike receptors, most of the molecule is on the outside of the cell. Differential stickiness of cadherins among the cells is what gives the blastula its shape. There are a wide variety of cadherin family members, and each sticks only to its own kind. In addition to about 20 different cadherin genes, each gene has multiple splice variants, in which particular exons are skipped or included in the production of the protein (see section 6.1), giving rise to hundreds of different cadherins. Particular cadherin genes are often organ specific, and the splice variants appear to come and go at different developmental stages.

The next developmental stage is called **gastrulation**, during which an even more structured set of cells, called a **gastrula**, is formed. A fundamental division of cell types happens at this stage, as well as large-scale movement that rearranges the positions of these types. In the Bilateria, gastrulation forms three distinct types of cells, called **germ layers**: the **ectoderm** (Latin for "outer skin") the **endoderm** ("inner skin"), and the **mesoderm** ("middle skin"). Additionally, germline cells (which will end up producing sperm or eggs) segregate at this stage.

The formation of these layers is a remnant of a very early adaptation in multicellular animals. Recall that *Volvox* and Dictyostelium use very simple position signals to control the differentiation of their two cell types. The next step in the evolution of multicellularity, the origin of these germ layers, is illustrated by hydra, a metazoan that split off before the origin of the Bilateria and is shown in figure 7.3. Hydra have two of these layers, an endodermal gut and an ectodermal skin, and they roll up to form tubes somewhat reminiscent of an embryo after gastrulation. Randomly dispersed in the tubes are cells that contract rhythmically, allowing the hydra to move rapidly through the water

Figure 7.3
Two hydra, one large and one small. These simple multicellular organisms form two-layer, tube-like bodies. The tubes are able to contract rhythmically, allowing the hydra to rapidly move through water gathering food. Photograph courtesy Hans Bode, and is used with his permission.

sucking up algae, a very impressive adaptation. The oldest Bilateria, such as flatworms, confined these musclelike cells to a new layer, the mesoderm, and the three-layered body plan was born.

The cells in each of these layers have different fates, which are preserved throughout development. The ectoderm emerges first, and defines the outermost layer of the organism. In vertebrates, the skin, tooth enamel, the lining of the mouth, the eye, and the cells that will form the nervous system all arise from ectoderm. The endoderm is created when a set of cells pinches inward from the hollow ball, and migrates as a whole to form a sort of second ball inside the first. This inner layer in the gastrula will ultimately form the gut, as well as the linings of all the organs that connect to the gut, such as the liver and pancreas (see below). Endoderm also forms the air cells of the lungs.

Mesodermal cells break off from the gastrula and migrate in between the ectoderm and endoderm layers. These cells create a tough boundary membrane called a **peritoneum** that encloses the **coelom**, a fluid-filled body cavity within the abdomen, found in all Bilateria. This organization allows organs to be attached to each other, but still move freely within that cavity. These cells also produce a physical support network for the organs, made of structural proteins such as **collagen**, **laminin**, and **fibronectin**. These proteins form sheets, fibers, and gels of various kinds, collectively known as the **extracellular matrix** or **ECM**. The ECM provides structure and support for the growing organism, as well as paths for migrating cells to follow. Finally, mesodermal cells also specialize into most of the contents of the body that aren't skin or gut: for example, the bones, muscles, heart, circulatory system, and reproductive system. All organs include at least some cells, called **mesenchyme**, that arose from the mesodermal layer. Signals from the mesenchyme play an important role in directing the specialization of nearby endodermal cells, a process called mesenchymal patterning.

The motion that creates these layers is dramatic, and is only the beginning of an increasingly detailed cellular specialization and movement of tissue. Three distinct mechanisms underlie **morphogenetic movement** throughout development: **invagination**, a motion in which a set of cells moves inward, somewhat like an indentation made in a balloon with your finger. Invagination is the type of motion that drives the creation of the endoderm. **Ingression** occurs when cells drop the attachments to their neighbors and move as individuals. The original neighboring cells reattach to each other, and the migratory cells follow morphogenic signal gradients to their new homes. Ingression is how the mesoderm is formed. The third kind of motion, called **involution**, is when a sheet of tissue pushes over another, forming folded layers.

Cells can move long distances in the developing embryo. For example, all the nerve cells in the body arise from a small piece of the ectoderm called the **neural crest**. These cells will disperse by ingressing throughout the entire organism, forming the network of nerves that supports sensation and action.

7.2.4   Differentiation

Animal bodies involve the complex structural and functional coordination of a very large number of cells, as many as tens of trillions in whales. Fortunately for our understanding of them, these enormous numbers of cells can be classified into a relatively small number of groups, called **cell types**. Different cell types coordinate with each other, forming complex tissues and organs.

Cell types were traditionally defined based on the microscopic structure and the functions of the cell, methods that have been supplemented with molecular characterizations in an increasing number of cases. The Cell Type Ontology[7] currently lists over 800 types of cells, covering all organisms. Only about 200 cell types are found in human bodies.[8]

The process by which a cell becomes a particular type is called **differentiation**, and, like morphogenesis, this process is driven by a highly conserved set of transcription factors and signal transduction pathways. These transcription factors are generically referred to as **growth factors**, and the individual ones often have names that indicate the type of cell or tissue it produces. For example, the **bone morphogenetic proteins** or BMPs drive the differentiation of various types of cells. In humans, there are 16 known BMPs, causing differentiation in bone, teeth, cartilage, and so on.

The organization of the signals that result in cellular differentiation cannot be solely spatial, since multiple types of cells must be generated in one place to form an organ or tissue. In many cases this is supported by cell migration. Differentiated cells arise in a particular place and then migrate to multiple targets, where they coordinate with other cells to form more complex structures. Migrating cells can follow the same morphogen gradients that define location, and in some cases, target cells excrete migration signals that enable other cells to find them. Cell adhesion protein interactions then organize the various cell types into specific arrangements, leading to coordinated cellular structures involving many precisely arranged cells of varying types.

---

7. http://obofoundry.org/cgi-bin/detail.cgi?id=cell.

8. This is the currently accepted number as defined by differences that can be seen through a microscope. Molecular characterizations may dramatically increase the number of types of cells that can be recognized.

The need for new cells of a particular type occurs not only in embryos, but also throughout the life of an organism. For example, human red blood cells live for only a few days and need to be replaced constantly. The cells whose function it is to produce other cells of a particular type are called **stem cells**. Embryonic stem cells give rise to all of the cells of each cell type necessary for the developing organism. Adult stem cells divide throughout the lifespan of an organism to replace cells of a particular type, based on cell signaling cues. Stem cells are also important because many important cell types become **terminally differentiated**, meaning that they themselves cannot divide. Cell division is a disruptive process that stops most other processing and remodels the structure of a cell; recall that Volvocales could not both swim and divide at the same time. Many cellular specializations, such as those seen in neurons and certain muscle cells, are simply incompatible with cell division. These cells must be produced by stem cells, rather than dividing themselves.

Stem cells actually give rise to all of the cells in an organism, including other stem cells. The process of differentiation happens in stages, producing a series of increasingly specialized stem cells. The cells present in the morula have the potential to become any cell type, and are therefore said to be **totipotent** or **pluripotent**.[9] Pluripotent embryonic stem cells have great potential for regenerative medicine, but, as discussed in more detail in section 11.1, their use is controversial.

By the time of gastrulation, the cells that have differentiated into endoderm, mesoderm, and ectoderm no longer have the potential to become any sort of cell. They become **committed** to producing only cells of a certain set of types. Additional stages of differentiation further and further restrict the types of cells that can be produced, leading ultimately to **unipotent** stem cells, which produce only a single type of cell. The path from undifferentiated progenitor to a fully differentiated cell is called the cell type's **lineage**. Related cell types share portions of this lineage. For example, in adult humans, the bone marrow contains **hematopoietic stem cells** that are capable of producing all of the various kinds of red and white blood cells. Some of those stem cells produce **erythroblasts**, which are unipotent stem cells that produce only red blood cells.

9. Totipotent means able to give rise to any cell type. The first differentiation in humans (and other placental mammals) separates the cells that will give rise to the placenta and those that will give rise to all of the cells of the offspring. The cells that can give rise to any cell in the offspring, but not placental cells, are called **pluripotent**. Depending on context, pluripotent can also mean capable of producing more than one cell type, such as the hematopoietic stem cell discussed in the next paragraph.

The process of differentiation at a molecular level involves a switch like the one Volvocales uses. For stem cells, the switch has to make cell division result in two types of cells: one that takes on the specialized function of the differentiated cell type, and one that remains as a stem cell able to continue to produce more differentiated cells. The switch is initially thrown by differential concentrations of transcription factors in the cells resulting from the division (sometimes regulated by size differences in the daughter cells, just the way Volvocales does it). Changes in the conformation or methylation status of DNA (see section 6.1) in the differentiated cell can make the genes in those regions inaccessible to the transcription machinery. These changes ensure that differentiation is permanent; in normally functioning organisms, cells never de-differentiate. It is likely that this fact is part of the system an organism uses to ensure that selfish mutants cannot harm it. As discussed further in section 9.6, cellular de-differentiation is a hallmark of cancer.

7.2.5 Organogenesis

Once the three layers are formed, **organogenesis**, the phase of embryonic development after gastrulation, begins in earnest. The process that creates organs from differentiated cells is a remarkable dance that exploits multiple morphogenic gradients, complex cell-to-cell signaling, cell movement, and differential adhesion and it even takes advantage of physical factors like fluid pressure. These processes are still incompletely understood for many organs; cardiovascular development is perhaps the best worked out so far.

The development of the cardiovascular system starts very early in embryogenesis, in part since establishing a circulatory system dramatically improves the ability of a developing embryo to distribute nutrition and remove wastes, compared with simple diffusion. The embryonic circulatory system is quite different from that of the adult (for example, getting nutrients from a yolk or placenta rather than a gut), and it has to both function and develop at the same time.

All of the mechanisms discussed so far play a role in the development of the heart. Cardiac progenitor cells arise from the mesoderm, and differentiate into **endocardium**, which lines the valves and the interior of the heart, and **myocardium**, the heart muscle. Growth factors secreted by cells in the endoderm and Hox-based position-specific transcription factors begin this process of cellular differentiation in two locations, one on the left and one on the right of the embryo.

These two types of cardiac precursor cells, each arising from two different locations, migrate toward the midline of the organism, each following gradients of fibronectin in the extracellular matrix. Each of the migrating groups of cells has the ability to specify a complete heart; if they are prevented from joining, the organism will grow two hearts. The cells then join and adhere to each other based on the expression of a cadherin protein, forming a tube with endocardial cells inside myocardial cells. Nearby cells secrete inhibitory signals that prevent the formation of cardiac tissue in inappropriate places. The tube begins to differentiate based on transcription factors that are asymmetrically expressed from one end to the other, leading to the formation of the basic regions and chambers of the heart. The next stage, called looping, bends the tube into a sort of U-shape, and the cell types differentiate further, based on a new axis of local morphogens that defines adult heart position. Finally, the chambers and valves are formed and the heart fuses with the separately developed systems of arteries, veins, and blood. In humans, the last step in this process occurs at birth, when the circulatory system has to rapidly grow tissue to plug holes that connected the left and right chambers, in order to manage the transition to a baby breathing on its own for the first time.

The growth of the vasculature is also a remarkable process. The blood vessels that line the organs develop with the organ. The circulatory system develops separately, starting by developing a network of fine tubes, called **capillaries**, which then grow, are remodeled and pruned, and finally differentiate and connect up to the vessels growing in each organ. The exact connections that are formed in this complex system are determined by cell-to-cell signaling, and have a random component to them. No two people have exactly the same circulatory system, not even identical twins.

The development and interconnection of organ systems is complex, and the molecular details are still incompletely worked out. The exact functions of many members of gene families of growth factors, adhesion, and signaling molecules remain unknown, and the molecular mechanisms that underlie many embryonic events are in many cases just beginning to be elucidated. The powerful molecular tools now available, and insights from the multispecies evo-devo approach suggest that great progress in these areas is likely soon.

7.3   Growth

In many circumstances, larger size has a reproductive advantage. Increased size can reduce the threat of predators, and make gathering food easier. Many enormous organisms have been quite successful, from dinosaurs to blue whales (which, incidentally, are much larger than any dinosaur). Even organisms that people consider small (say, flies or minnows) are much larger than unicellular and early multicellular organisms, as reflected in Volvocales or Dictyostelium. However, with increasing size comes a variety of challenges. For example, increasing size changes the relationship between volume and surface area; transport of material within an organism becomes a pressing problem.

How did size evolve? Growth and development are often thought of as intimately related, but it now appears they are quite separate processes and they do not even generally occur at the same time. Conservation of matter ensures that an embryo can't grow in size without an input of new mass. Most animal embryos have no way of feeding, so they can't gain mass until they develop a gut, which requires significant time. Also, some of the morphogens that lay out the body plan spread by diffusion, which sharply limits size, at least of the early embryo. For these reasons, morphogenesis and cellular differentiation precede growth in nearly all animals, and this growth program must have arisen early in the history of animals. Additional evidence for this conserved growth program is that at the end of gastrulation, the embryos of most animals (say, fruit flies, fish, people, and whales) are about the same size, roughly a millimeter long, when they begin to elongate along the anterior-posterior body axis.

Growth in size, though seemingly a simpler process than laying out a body plan, is itself quite complex. One of the interesting aspects of the control of growth is its precision. Note that the growth of the two sides of the body in Bilateria produces parts that are almost exactly the same size, despite their distance from each other in the developing embryo. Many evolutionary adaptations (such as the elephant's trunk or the aye-aye's finger) involve precise changes in the size of one body part, independently of all of the others. Growth is regulated by molecular mechanisms similar to the morphogens. A large group of proteins, called **the transforming growth factor beta *(*TGF-β*)* superfamily**, includes many proteins that regulate growth (as well as some important morphogens). These are intercellular signaling proteins; their receptors are called **SMADs**.

An organism can grow in two ways: it can create more cells, called **proliferation**, or it can make the cells it has bigger. Both mechanisms are widely used, either together or at different times, and each has fundamental limitations. Proliferation doesn't always result in growth; it is often the case that cells divide without growing, producing more, but smaller, cells. There is also a tension between proliferation and differentiation; terminally differentiated cells cannot divide at all. Limitations on the size that a cell can grow appear to arise based on the need for cytoplasmic signals to reach the nucleus. Cells that grow to a very large size often have multiple nuclei, having replicated their DNA but not divided.

The knowledge of the complete developmental program of the worm *C. elegans* illustrates some of the approaches that animals take to address these limitations. Starting from a large egg, embryogenesis and differentiation produce a nearly fully formed organism of hundreds of cells, but it is not much bigger than the single egg cell. After hatching, the worms are voracious eaters, and grow about 100-fold in size, mostly through increases in the size of their cells without further cell division. Some cells, particularly those in the gut of the worm, replicate their DNA without dividing as they grow, ultimately producing very large cells with 32 nuclei each. A few types of cells, particularly those involved in reproduction, continue to divide in the adult worm. Most of those cells remain relatively undifferentiated until quite late in life, illustrating the tradeoff between differentiation and proliferation.

Most animals have to pack all the material necessary to produce an offspring into an egg, and growth is limited until the egg hatches. The simplest and most widespread approach to dealing with these constraints is to produce a big egg. Many organisms produce very large egg cells (for example, both frog and fly eggs are about 100,000 times larger than their other cells). These large eggs undergo embryonic development without much growth, and only after birth does a hatchling grow through feeding-driven increases in cell size.

Growth driven by increases in cell size is limited, even with mechanisms like creating multinucleated cells. One approach to these constraints that many animals have taken is to insert a larval stage between embryonic development and the formation of an adult. A larva is capable of feeding, and most larvae grow significantly. Larvae carry around sets of undifferentiated cells called **primordia**, which will eventually develop into adult structures. These cell populations increase in number as the larva grows by feeding, but only differentiate and unfold their morphogenic program to produce adult structures later, often at an explicit transition from larva to adult stages. At that transition (which often takes place in a protective casing that prevents further feeding or growth), the larval tissues are broken down and their contents are used to support the growth of the primordia, which then differentiate into adult tissues. Some animals go through multiple juvenile stages before becoming adults. This larval lifestyle is an adaptation that was able to produce much larger organisms than the big egg strategy alone.

Vertebrates grow much larger than other organisms, and the basis of that increased size is expansion in the use of egg yolk, a food supply that a developing embryo can use without first developing a digestive tract. Some insects also produce eggs with yolk, but vertebrates have dramatically more yolk than other organisms. The eggs of jawless and cartilaginous fish, useful proxies for the early ancestors of all vertebrates, are enormous relative to invertebrates, some reaching more than an inch in diameter. The egg cell itself is not nearly so large, but the volume of yolk produced by the mother allows growth to proceed as if it were. Large yolks support a much more extensive period of growth in the developing embryo, allowing the production of much larger embryos, and ones with more different cell types.

Organisms do not have to grow at uniform rates in all tissues. In fact, different growth rates in different tissues during development can result in coherent, large-scale changes in the final form of the organism, a process called **allometry**. For example, some rodents, like mice, have seed pouches inside of their mouths; others, like pocket gophers, have external, fur-lined seed pouches. The difference is a result of a change in the rate of growth in the embryonic precursors for the pouches relative to the rest of the snout. The different growth rates lead to changes in the relative position of the tissues. Once the position changes, other changes, like the growth of fur in the pouches, arise because of the morphogens in the new neighborhood.[10]

The most recent step in the evolution of fully terrestrial animals was the development of an egg that didn't require being laid in water, a major evolutionary transition that opened enormous new niches for exploitation. These eggs depend on the inclusion of amniotic fluid and a complex system of membranes to protect the egg from drying out, again driving up the extent of maternal contribution to embryonic growth. The placental mammals, including people, take this contribution even further, providing direct maternal support of embryonic and fetal nutrition, which makes possible very long periods of growth before birth.

7.4 Multicellular Life and Humanity

Most of the material covered up to this point is universal, or nearly so, among eukaryotes or metazoans. The molecular bases of development, gene regulation, protein production, DNA replication, and so on are the same in people as they are in mice, worms, and flies. It is these molecular homologies, in both structure and function, that have made it possible to understand so much about the biology that so many creatures share.

The preceding discussion of growth focused on the general problem of becoming big, but the particular mechanism responsible for human embryonic growth appeared only in the last sentence, and without much detail. Uterine support of embryonic growth occurs only in a few thousand species, far less than 1% of the million or so known species of animals. Other growth mecha-

10. See, for example, the discussion of modularity in *Developmental biology*, by Scott Gilbert (Sinauer Associates, 2006).

nisms are more widespread, and more ancient. Since our heritage includes those mechanisms, it is likely that they remain within us, adapted perhaps to different or more restricted functions.

Nevertheless, biological understanding of the particular adaptations of our own species matters a great deal to us. More recently evolved functions, such as uterine development, and especially unique human capacities, such as language, are of particular interest. These recently evolved capacities are generated by quite small differences in DNA. Many protein coding sequences are quite highly conserved among all vertebrates, and variation within the mammals (including humans, mice, rats, dogs, cats, monkeys, and so on) is even smaller. Most of the genetic differences among mammals are in transcription factor binding and other regulatory sites (e.g., changing development), not so much in the proteins themselves.

These relatively small molecular differences, and the relatively smaller numbers of other species available for comparative studies make scientific progress more difficult. The molecular differences may be small, but the consequences for human health and well-being can be large. Consider the case of muscular dystrophy, a devastating genetic disease. The gene that is mutated in the disease, called dystrophin, was one of the first identified with the techniques of contemporary molecular biology. Yet progress on treating the disease remains poor, since that protein apparently plays a different role in humans than it does even in most other mammals. Creating a similarly mutated version of the mouse homolog of that gene has very little effect on the mouse, making it hard to study the disease mechanism in the laboratory. Dogs are affected in much the same way people are, but their slow development (compared to mice) and other differences makes it harder to use them to develop therapies.

Beginning in the next chapter, the focus will turn to the issues that are of importance to human beings, highlighting homologies whenever possible. Our anatomy and physiology shares much with other organisms, particularly other mammals, but there are also clearly aspects of our biology that are unique. Similarly, the diseases that plague human beings sometimes, but not always, affect other organisms as well. Fortunately for our knowledge of human biology, most of the features that shape us are widely shared, and even the aspects of us that are clearly novel (such as language) have increasingly well understood molecular evolutionary roots.

7.5   Suggested Readings

The excitement of the evo-devo approach has spawned a good number of popular texts on molecular developmental biology, starting with Sean Carroll's outstanding *Endless Forms Most Beautiful: The New Science of Evo-Devo*.

Christiane Nusslein-Volhard, who won the Nobel prize for first identifying genes that played a role in development, wrote *Coming to Life: How Genes Drive Development* for popular audiences. Her perspective does not include the evolutionary angle in evo-devo, but still makes a compelling read. Debra Niehoff's *The Language of Life: How Cells Communicate in Health and Disease* is a very accessible treatment of cell signaling in many contexts. Eric Davidson's *The Regulatory Genome: Gene Regulatory Networks in Development and Evolution*, is a more difficult and technical text, but rewards the extra effort with a deep understanding of how molecules shape bodies.

Developmental biology is a vast subject, with many good textbooks. Probably the most widely used of these is Scott Gilbert's *Developmental Biology*. A more molecular text is *Molecular Principles of Animal Development*, by Arias and Stewart.

8 Anatomy, Physiology, and Systems Biology

Knowledge of the structures and functions of the molecular components of life underpins the understanding of whole organisms. However, biological wholes have functional capacities that are not apparent in their disconnected parts. Assemblies can produce **emergent properties**, particularly if there are complex relationships among the parts or between the whole and its physical, chemical, and biological environment. For example, cardiomyocytes (heart cells) contract rhythmically. That property is not caused by some rhythm-generating molecular part, but instead emerges from the interplay of multiple ion channel proteins, time-varying ion concentrations inside and outside of the cell, and the role of those ions in the functioning of contractile proteins. Understanding the structure and function of body parts requires explanations that integrate many levels and types of phenomena.

Anatomy is the study of the structure of living things, and is often divided into **gross anatomy**, the study of organs and structures than can be seen by the naked eye; **histology**, the study of the organization of cells into tissues; and **cytology**, the study of the structure of individual types of cells. **Physiology** is the study of the functions of cells, tissues, organs, and bodies, with an emphasis on how parts interact to work as integrated wholes, and on how the many levels of organization influence each other. The allied field of **systems biology** takes a similarly integrative perspective, focusing on interactions at the molecular level.

8.1 Homeostasis

The basic function of a multicellular body is the same as that of any other living thing: reproducing. Since multicellular organisms must go through a time-consuming process of development, staying alive long enough to reproduce is an even greater challenge for them than it is for unicellular organisms.

The advantages of multicellularity arise from cellular specialization, but specialization also generally involves the loss of some other functionality. Multicellular organisms must therefore create an internal environment that ensures that all their specialized cells get what they need to survive. The process of keeping all the cells in a body in a supportive internal environment is called **homeostasis**.

The internal liquid environment of a multicellular body that bathes all the cells is called the **interstitial fluid**. Homeostasis is largely about regulating the state of the interstitial fluid, keeping many of its parameters, such as temperature, pressure, pH, and so on, within the ranges needed by the body's cells. Concentrations of many small molecules, both nutrients (e.g., glucose or oxygen) and toxins (e.g., carbon dioxide or excess nitrogen), need to be tightly regulated throughout the body. Maintaining the concentrations of the ions needed for life (including sodium, potassium, calcium, magnesium, and phosphate), collectively called **electrolytes**, is similarly vital.

Interstitial fluid contacts all the cells of an organism, and all the organ systems can be seen as working to regulate its parameters. The circulatory system communicates with the interstitial fluid through both diffusion and active transport. Blood flow is critical to ensuring that the proper range of concentrations of nutrients and toxins are maintained throughout the interstitial fluid, particularly in large organisms. Other systems tend to interact with the blood to make their contribution to homeostasis.

Homeostatic regulation often involves **negative feedback loops**, where receptors detect deviations from target ranges and communicate with effectors that take actions to move in the opposite direction of the deviation. For example, the level of glucose in the blood needs to be maintained within a certain range, but demand for ATP (and therefore glucose) is unpredictable. **Beta cells**, a part of the **pancreas**, contain a glucose-sensing signaling pathway (the first enzyme of which is a hexokinase, a homolog of the enzyme that starts glycolysis). When glucose levels fall below the target range, say because of high demand for ATP in muscles, beta cells excrete a hormone called **glucagon** into the bloodstream. Glucagon receptors on cells in the liver detect this message, and adjust their metabolic pathways to convert the energy storage molecule glycogen (see section 3.3) back into glucose and secrete it into the bloodstream. When glucose levels exceed the target range, say after a nice meal, the beta cells instead produce insulin, which drives liver, muscle, and other cells to import glucose from the blood and convert it into glycogen. The balance of glucagon and insulin creates a negative feedback loop that tightly regulates blood sugar levels.

However, homeostasis is not just a simple set of negative feedback loops that lock physiological parameters within a particular range. Many interacting layers of control allow the organism to dynamically respond to its environment in the manner that is most likely to keep it alive. Consider the effect of the hormone adrenaline, which is released when the organism detects an important external threat. Adrenaline overrides the insulin/glucagon signaling mechanism, ensuring the liver generates as much glucose as possible—in anticipation of increasing energy requirements to deal with the threat. Homeostasis is not a passive process, but finely tuned control of the entire system, resisting all varieties of threats and stresses, both internal and external.

The regulation of glucose levels involves more than just conversion into and out of glycogen; there are other, longer-term energy storage mechanisms as well. Prolonged excesses or deficits of glucose can trigger a number of additional mechanisms involving feeding, fat storage, and other responses to starvation or plenty. The presence of insulin implies that energy is plentiful, and it has many effects beyond influencing glycogen production, such as altering cellular metabolism, growth, and even cell division in dozens of ways. Insulin receptors are found in nearly all cells in the body, but trigger quite different responses in different types of cells.

All the systems in the body must interact to maintain homeostasis. Anatomically, bodies are organized into **organs**, which physiologically function together as **organ systems**. Since the organ systems all interact with each other, the definition of boundaries is to a certain degree arbitrary, but animals are generally said to have ten different organ systems: The **circulatory system** transports nutrients (and toxins) via the blood, heart, and vasculature. The **digestive system**, which includes the esophagus, stomach, intestines, liver, gallbladder, and pancreas, extracts nutrients from food and regulates the levels of those nutrients in the interstitial fluid. The **endocrine system** includes a variety of glands that secrete hormones into the blood that function to regulate blood pressure, ion levels, metabolism, and other critical aspects of homeostasis, as well as to produce periodic changes such as circadian rhythms and menstrual cycles. The **immune system** defends the body against invading organisms and, sometimes, aspects of the self that have become damaged or cancerous; it includes the white blood cells, spleen, thymus, and its own circulatory system of lymph nodes and vessels. The **integumentary system** includes skin, hair, feathers, claws, and teeth, and supports a variety of homeostatic functions, including temperature regulation, secretion of various substances, sensation, protection from the environment, and, in insects and some reptiles, respiration. The **musculoskeletal system** enables an organism to move, perhaps the most central adaptation of the animals, as well as providing the solid structure that defines body morphology. In most animals, the muscles and bones are the largest organ system, making up more than half of body mass. The **nervous system** senses the environment, integrates information, evaluates situations, coordinates motion, and exerts moment-to-moment control over a wide variety of homeostatic parameters. The structures of the nervous system support an extraordinary ability to change (called **plasticity**), underlying its key adaptation: learning from experience. The **reproductive system** produces gametes, regulates fertilization, and gestates embryos. This system is central to the existence of the organism, and has powerful influence over all homeostatic parameters. The **respiratory system**, which in land animals includes the nose, trachea, lungs, and diaphragm, functions to regulate gases dissolved in the interstitial fluid, primarily oxygen and carbon dioxide. The **urinary system** includes the kidneys, ureters, bladder, and urethra, and functions to remove surplus and waste molecules; it can also concentrate certain ions when supplies are low.

Although differences in the organ systems are found in different animals, there is also many commonalities. Each of these systems is increasingly well understood, from molecular levels to anatomic and physiological function. These systems can become quite complex, particularly in vertebrates; however, the organs that constitute them are composed of various combinations of a modest number of tissues.

8.2 Tissues

The first level of biological organization above the cell is the **tissue**, a collection of cells of particular types in a particular spatial distribution. Plants have three basic tissue types: epidermis, which forms the outer surface of the leaves and plant body; vascular tissue, which transports fluid and nutrients internally; and ground tissue, which is everything else, including photosynthetic and energy storage cells. In animals, there are four basic types of tissues: epithelium, muscle tissue, nervous tissue, and connective tissue. The focus here is on the tissues found in animals.

Epithelium is boundary-forming tissue. The surface of the skin is made of epithelial tissue, as are the linings of all of the organs of the body: the lungs, the gut, blood vessels, and so forth. Epithelial cells also form **glands**, which produce and secrete hormones, the global signaling molecules. In the preceding example of glucose regulation, the beta cells are epithelial cells, insulin and glucagon are hormones, and the pancreas is a glandular organ.

There are dozens of epithelial cell types, classified by their shape, layering, and particular specializations, as illustrated in figure 8.1. Epithelial cell shape can be cuboidal (in the shape of a cube), columnar (taller than they are wide), or squamous (flattened, like fish scales). Squamous epithelium is found in places like the skin, lung, kidney, and the lining of blood vessels, where it manages the diffusion of water, ions, or other materials across boundaries. Columnar epithelium, such as the tissue that lines glands or the intestines, is usually involved in the absorption or excretion of more complex materials, such as proteins or hormones. These tissues can be composed of a single layer of cells, called simple epithelium, or multiple layers anchored by a **basement membrane**, called stratified epithelium. Stratified epithelium is often found in tissues subjected to external stress, such as the skin or intestines. Transitional layering is found in epithelial tissue that stretches, such as in the bladder; the cells appear to be stratified when contracted, but can slide over each other when stretched to form a single layer. Some kinds of epithelial cells have particular specializations: Keratinized epithelium, found mainly in the skin and characterized by large amounts of the protein **keratin**, forms a tough, nearly impenetrable layer. Ciliated epithelial cells have little hairlike appendages, called **cilia**, on their exposed surfaces. The cilia beat rhythmically and function to move substances, such as mucus, through ducts.

Connective tissue involves a diverse group of structures characterized primarily by the importance of the material outside of the cells, called the extracellular matrix, or ECM. Connective tissue creates the structural components that hold organs together and in place in the body. Proper connective tissue includes fat, ligaments, and tendons, which actually do connect structures to each other. The ECM produced by this connective tissue is largely fibrous, and is produced by cells called **fibroblasts** (the suffix **blast** indicates a cell type produces something: fibroblasts make fibers). Some connective tissue makes gooey, often adhesive ECM called ground substance. Proper connective tissue is divided into loose, meaning lots of ECM and relatively few cells, and dense, which has more cells and less ECM. Elastic connective tissue, such as that surrounding blood vessels or making the disks between vertebrae in vertebrates, can stretch and contract. Bones and cartilage are called supportive, rather than proper, connective tissue. Blood and lymph, where the ECM is liquid, are called reticular connective tissues.

Muscle tissue supports only one function: contraction. It produces force and causes motion, both of the entire organism (called locomotion) or within an organism (e.g., to pump blood or push food through the gut). The main molecular structures that underlie contractile function are the proteins actin and **myosin**. Actin is the central component of the cytoskeleton of all

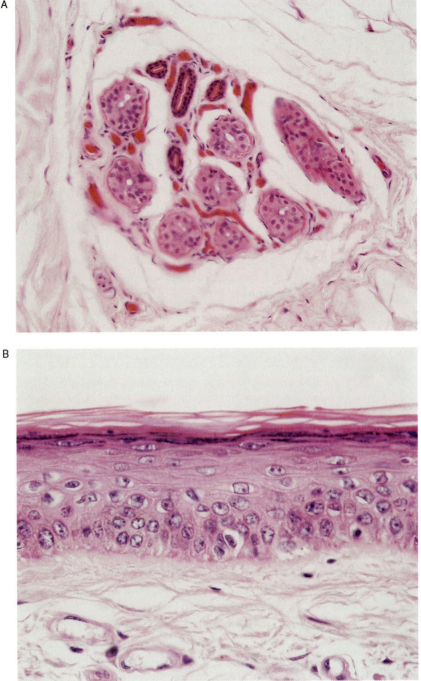

Figure 8.1
Human epithelial cells. Panel (A) shows stratified cuboidal epithelium in sweat glands from skin. Panel (B) shows an image from a lip; the upper half of the image is composed of stratified squamous epithelium.

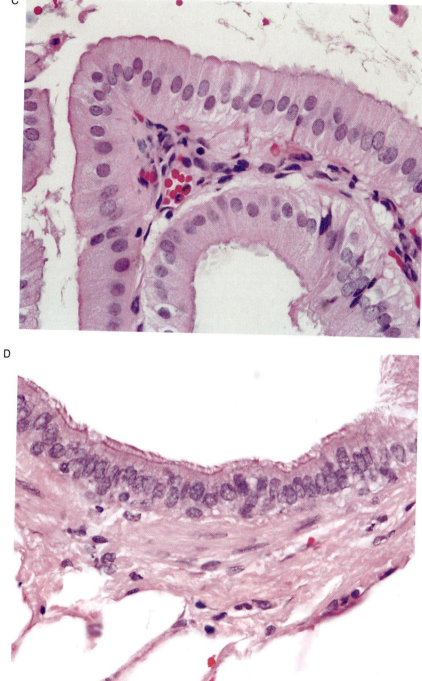

Figure 8.1
Panel (C) shows simple columnar epithelium from a gallbladder. The top layer of tissue in panel (D) is ciliated simple columnar epithelium from the lung. These images are from the excellent Blue Histology Web site at http://www.lab.anhb.uwa.edu.au/mb140/ and are copyrighted by Lutz Slomianka, and used with permission.

eukaryotic cells and was discussed in section 6.2. The contractile ability of muscle cells arises from the ability of myosin to ratchet itself very rapidly down an actin fiber, dramatically changing the shape of the cell, as illustrated in figure 8.2. Muscle cells are **excitable**, meaning that they are triggered to contract by changes in the electrical potential (i.e., ion concentrations) across their membranes. By working together in enormous numbers, myosin molecules are able to generate macroscopic forces. More than 10 trillion myosin molecules are needed to generate enough force to hold up 1 kilogram. Each one requires ATP to fuel its activity, which is the reason muscle activity is such a prodigious consumer of energy.

There are three types of muscle tissue: cardiac, smooth, and skeletal muscle. Cardiac muscle forms the heart, the center of the circulatory system, and has a variety of special properties. Cardiac muscle contracts spontaneously (without

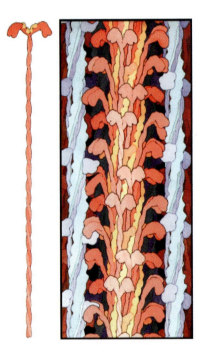

Figure 8.2
On the left, a single, long-tailed myosin dimer, consisting of two copies each of light (lighter orange) and heavy (darker orange) myosin chains. On the right, the molecular structure of a myosin thick filament, consisting of about 300 myosin molecules whose heads are making contact with the actin fibers, shown in blue. Powered by ATP, the heads flex and snap downward, moving the entire assembly a step at a time along the actin molecule. These images were created by David S. Goodsell of the Scripps Research Institute, and are available from the wonderful PDB Molecule of the Month Web site (http://www.rcsb.org/pdb/static.do?p=education_discussion/molecule_of _the_month/).

any external signal) and rhythmically. As described briefly earlier, beating arises through the activity of several sorts of ion channel proteins in the membranes of these cells. These channel proteins are **voltage-gated**, which means that they open and close depending on the voltage across the membrane. The voltage across the membrane is a measure of a difference in charge, which is determined by the concentrations of various ions on either side of the membrane. The change in ionic concentration driven by the opening and closing of voltage-gated ion channels builds up over time, changing the voltage across the membrane. When the change in voltage becomes large enough, it causes the channel proteins to change state, reversing the ion flows. Over time the voltage drops, bringing the cell back to the original state. Feedback with delay causes oscillation, making these cells beat.

One of these ions, calcium, is central in activating myosin, and passing an internal threshold calcium level induces a contraction. This process causes a single cardiomyocyte in an appropriate ionic environment to beat rhythmically. Cardiomyocytes not only beat spontaneously, they coordinate spontaneously as well. Cardiac muscle cells are connected to one another through **gap junctions**, a protein pore that allows certain molecules and ions to flow freely between the cells. This linkage allows cardiac muscle cells to beat together. In the intact heart, waves of calcium ion channel activation (and influxes of calcium ions) start from a small set of pacemaker cells and spread at high speed to create a powerful rhythmic contraction at a macroscopic level. Cardiac muscle has a large number of mitochondria and is not susceptible to tiring, but it does require a constant supply of oxygen and glucose to do its work.

Skeletal muscle is the sort of tissue most people think of as muscle; it is responsible for locomotion and body posture. Skeletal muscle cells, also called muscle fibers, are large and contain multiple nuclei, as shown in figure 8.3. The nuclei are pushed to the edges of the cylindrical cells, leaving the center of the cell to be packed densely with **myofibrils**, a special organelle that contains actin and myosin fibers and runs from one end of the cell to the other, with attachments at the membrane. Muscle tissue is organized into bundles called **fascicles**, which are surrounded by connective tissue. Contraction of skeletal muscles is triggered by nerve impulses.

Smooth muscle, which lines the walls of blood vessels and other organs (and keeps a mollusk's shell closed), is cytologically different in appearance from other muscles. It is not under voluntary control, and has adaptations like those in cardiac muscle to ensure it doesn't tire.

Nervous tissue underlies sensation, thought, voluntary action, and much of the regulation of the other organ systems. **Neurons** are excitable cells that integrate and communicate information through rapid changes in ion

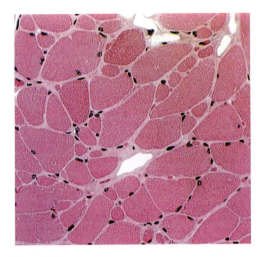

Figure 8.3
A photomicrograph of skeletal muscle cells. The cells are densely packed with myofibrils (stained
pink), and their nuclei (dark spots) are pushed to the edges. This photograph was taken from
Thomas Caceci's outstanding Web site on veterinary histology at http://education.vetmed.vt.edu/
curriculum/VM8054/VM8054HP.HTM and is used with his permission.

concentrations (and hence voltage) across their membranes, called **action
potentials**. Voltage-sensitive membrane channel proteins cause the change in
voltage of an action potential to move rapidly along the neuron's membrane.
An action potential travels down a neuron at between 10 and 100 meters per
second. Many vertebrate neurons are covered in **myelin**, an electrically insu-
lating phospholipid layer that dramatically increases the speed of propagation
of an action potential. Action potentials travel down **axons**, fibers that extend
from the neuron's cell body, transmitting information over long distances.
Axons from some nerve cells can be very long, for example reaching from the
base of the spine to the end of the foot. Nervous tissue also includes cells
called **glia** that surround the axons, producing myelin. Glia outnumber neurons
by about 10 to 1, and are increasingly recognized as having a significant
information processing role, although they are not themselves excitable.

   Sensory neurons, such as light-sensitive **photoreceptors**, or **proprioceptors**,
tors, which report on the relative position of body parts, respond to particular
changes in the state of the body or the external environment by firing action
potentials. **Motor neurons** send axons to all the muscles in the body. An action
potential in a motor neuron excites the muscle it controls through a **neuromuscular junction**, causing the muscle to contract. **Interneurons** link one
neuron to another, integrating sensory information into representations of the
state of the organism and the world, and triggering elaborate motor programs

that tend to achieve reproductive and survival goals. The properties of a nervous system are largely emergent, depending critically on how the constituent neurons are connected to each other, as described in section 8.3.3.

Tissues grow together in structured patterns to produce organs. Even muscles are not made only of muscle tissue; a muscle includes proper connective tissue that bundles the muscle tissue into fascicles, and attaches the muscle to the skeleton, as well as a densely integrated system of blood supply and enervation. Organs and organ systems organize tissues into the functional wholes that make up a body.

8.3   Organ Systems

Detailed description of vertebrate, particularly human, anatomy and physiology is readily available, and the treatment here is necessarily brief. Only the cardiovascular, immune, and nervous systems are discussed in this section. These are particularly important to human health, and convey some of the basic ideas of physiology.

8.3.1   The Cardiovascular System

The cardiovascular system transports most homeostatically regulated substances, including water, oxygen, proteins, hormones, glucose, ions, and various toxins. It can also play a role in the regulation of body temperature and pH. Some invertebrates have an open circulatory system, meaning that there is no distinction between blood and the interstitial fluid, and only a limited, open vasculature. Transport in these organisms is facilitated by muscle movement, including a beating heart. In some even simpler animals (e.g., platyhelminths, a sort of flatworm), no circulatory system is present at all, and transport is solely by means of diffusion; the digestive system of such animals must traverse the entire the organism, so no cell is too far from a source of nutrients.

The circulatory system of all vertebrates, and some invertebrates such as the annelids (e.g., earthworms) and cephalopods (e.g., squid) is closed, meaning that interstitial fluid is separated from blood, which is pumped by a heart through a system of vessels. Although the details of the structure and function of the circulatory system differ even between fairly closely related organisms (not even identical twins have exactly the same circulatory system; see section 7.2), the broad principles are highly conserved among these animals.

Blood is a connective tissue that consists of three cell classes suspended in a liquid called **plasma**. Plasma is mostly water, but also contains dissolved proteins, ions, hormones, and glucose, as well as waste products such as carbon dioxide and excess nitrogen. The three types of cells found in blood are red cells (**erythrocytes**), white cells (**leukocytes**), and **platelets** (sometimes called **thrombocytes**). The vast majority of the cells in blood are **red blood cells**, which make up a bit less than half of the volume of the blood (the measurement of the proportion of blood composed of erythrocytes is called hematocrit). Erythrocytes are packed nearly full with hemoglobin, a molecule that transports oxygen. Mature erythrocytes have no cell nucleus or organelles, and a relatively limited lifespan; in humans they live about four months. The white cells make up about 1% of blood volume, and are functionally part of the immune system (described later). Platelets make up less than 1% of the blood volume, and their role is to form blood clots when the circulatory system is damaged. Platelets are activated through exposure to collagen (indicating a breach in the vasculature) or through molecular signals such as the protein thrombin. Activated platelets send signals that activate other platelets, which aggregate, adhere to one another, and trigger a cascade of protein activity both within the platelets and in the plasma, ultimately producing a clot. All blood cells are produced by a series of increasingly differentiated hematopoietic stem cells in the bone marrow. The molecular understanding of the mechanisms of cellular differentiation arose largely from the study of this process.

Blood circulates throughout the body through a large system of vessels. Blood vessels are lined with a layer of simple squamous epithelium called the endothelium, which controls the flow of materials in and out of the vessel. Capillaries, the smallest blood vessels, are made of little but endothelium. Larger vessels surround the endothelium with connective tissue, vascular smooth muscle, and another layer of connective tissue that contains nerves and capillaries supplying the smooth muscle. That muscle regulates the diameter of the vessel (through **vasodilation** and **vasoconstriction**), which in turn changes blood flow in the system. Blood vessels are divided into **arteries**, which carry blood from the heart, and **veins**, which return blood to the heart. Capillaries connect the arteries to the veins, and perfuse the entire body. Although capillaries are very small, there are a very large number of them, about 600 in a typical cubic millimeter of tissue, each serving only about 500 cells. Some organs, such as brain, liver, kidney, and heart, have as many as 3,000 capillaries per cubic millimeter, ensuring that no cell is more than a few microns from the blood supply. Despite having the diameter of about single red blood cell, the cross-sectional area of all the capillaries added together is much larger than the cross-sectional area of the arteries and veins. As the same volume of blood is flowing through the larger cross-section, flow is slower and at lower pressure in capillaries than in arteries.

The cardiovascular system has hydraulic qualities that are unusual compared to the mechanical world of pipes and pumps. The vessels are as important in controlling blood flow as the heart: both the pressure and the circulation rate of the blood are influenced by vascular resistance (friction) and compliance (elasticity). The system is closed and under pressure. The total system pressure is influenced by fluid intake, kidney activity, and the compliance of the blood vessels. A heartbeat creates a pressure gradient between arteries and veins, resulting in blood circulation. The heart fills passively, unlike most pumps that both suck in and push out. For that reason the heart does not produce the pressure in the circulatory system, it only distributes the pressure.

Blood pressure varies significantly throughout the circulatory system. It is relatively high in the arteries, where the output of the pumping heart interacts with the elasticity and muscular control of the arteries to regulate arterial pressure. The pressure drops dramatically in the capillaries, because of their larger cross-sectional area. That area is so large that if all the capillaries were filled simultaneously, total pressure would drop dangerously. Instead, constriction and dilation of the vessels that connect arteries to the capillaries, called **arterioles**, specifically direct regional blood flows, playing a large role in the blood flow seen by particular organs. The veins act as a reservoir, holding the majority of blood in the body at relatively low pressure. Increased demand for blood (say, by muscles in need of oxygen) causes vasoconstriction in the veins, which reduces blood volume in the reservoir, ultimately increasing arterial volume. Blood pressure in various parts of the circulatory system and body parts are critical homeostatic parameters, and are controlled by a complex set of physiologic responses, pressure-sensing molecules (called **baroreceptors**), and hormonal signals, all under the direction of the nervous system.

Mammals actually have two linked circulatory systems, one that distributes blood to the body, called **systemic circulation**, and one that circulates blood through the lungs, called the **pulmonary circulation**. In the systemic circulation, blood in the arteries (leaving the heart) has more oxygen than blood in the veins (returning to the heart); in the pulmonary circulation, the situation is reversed, which can be confusing. These two systems generally have equal volumes of blood flowing through them. The maintenance of equilibrium between these two linked systems is a consequence of the passive filling of the heart.

The human heart has four chambers: two **atria**, which are filled by flow from systemic and pulmonary veins, and two **ventricles**, which pump blood out of the heart. The right atrium receives blood from the systemic veins and forces it into the right ventricle. The right ventricle then forces that blood into the pulmonary arteries. The left atrium receives oxygenated blood from the pulmonary veins, and forces it into the left ventricle. Finally, the left ventricle forces the oxygenated blood into the systemic arteries. The two atrial contractions (at the end of the **diastolic events**) occur simultaneously, as do the two ventricular ones (at the end of the **systolic events**), resulting in the familiar "lub-dub" sound of a beating heart. Heart valves, which are made of connective tissue, not muscle, ensure that the blood always flows in the proper direction.

The cardiovascular system is one of the most central to life. Disruptions to it, whether damage to the heart, disregulation of blood pressure, or loss of blood, are often rapidly fatal.

8.3.2   The Immune System

The free rider problem is an inherent part of evolving systems, so all organisms are confronted with a wide array of parasites and other invaders. Even bacteria have to deal with invaders, particularly viruses (bacterial viruses are called **phages**). This problem becomes more difficult for multicellular organisms with long lifespans, since the much shorter generation time of their pathogens means that the hosts are at an evolutionary disadvantage.

A critical aspect of any defensive system is the recognition of what is self versus what is nonself. Bacterial defenses are mostly based on DNA-cutting proteins, called **restriction enzymes**, that cut nonself DNA into pieces. How do bacteria ensure that their own DNA isn't cut up? Restriction enzymes are very specific, cutting only at a particular nucleotide sequence. Having restriction enzymes that cut at sequences not found in the bacterium's own genome is a basic way to distinguish self from nonself. However, increasing genome size diminishes the number of sequences that are not found in the self. Another mechanism, covalently adding a methyl group to the self DNA targeted by a restriction enzyme, protects the bacterium's own genome from being attacked. This **restriction-modification** system might have initially arisen from the bacterial DNA mismatch repair system, which also exploits DNA methylation patterns.

Unicellular eukaryotes developed another system for protecting against viruses, called RNA interference, or RNAi. As described in section 5.2.2, the RNAi system suppresses any messenger RNA that is complementary to a **double-stranded RNA (dsRNA)**. Some viruses invade in the form of dsRNA, which this mechanism completely silences. It is also possible for an organism to produce its own dsRNA (by coding for both the RNA and its complement), which leads to the degradation of any mRNA that matches that sequence. The dsRNA doesn't need to be a full-length copy; **small interfering RNAs** (siRNA) need only about 20 base pairs of overlap to be effective, making it possible to code defenses against certain invaders directly in the genome.

The protists also developed another system for detecting and repelling bacteria, called **antimicrobial peptides**. These short proteins (generally less than 50 amino acids) contain mostly hydrophobic residues surrounding two or more positively charged ones. These proteins interfere with bacteria through a variety of mechanisms, disrupting cell membranes, attacking the peptidoglycan wall, or causing damage in the cytoplasm. This approach to preventing bacterial infections is present in all eukaryotes, including protists and people; more than 900 different antimicrobial peptides have been identified.

The transition to multicellular organisms facilitated the development of specialized immune cells, but also complicated the task of discriminating between specialized self cells and actual invaders. Multicellular immune systems recognize invaders through both the presence of foreign materials (such as molecular characteristics of bacterial cell walls, or the presence of mannose, a sugar used primarily by bacteria) and the requirement that every cell in the body express a self-signal on its membrane.

Immune cells were one of the very early specializations, as evidenced by their presence in nearly all multicellular organisms, including, for example, sponges. These cells, called **phagocytes** (Latin for "eating cells") engulf and destroy invaders. Over time, a variety of even further specialized phagocytes developed. In vertebrates, many different white blood cells, such as **macrophages**, **monocytes**, and **neutrophils**, have as their primary function consuming invading cells. Bilateria had to address new kinds of invaders, such as fungi and protists, which didn't exhibit the bacterial molecular signals, and new aspects of the immune system arose to cope. For example, the **complement** system, a set of proteins present in the blood of both vertebrates and invertebrates, opens channels in the membranes of many kinds of invading cells (not just bacteria) causing them to swell with water and burst. **Natural killer** (or **NK**) cells recognize abnormal self cells, such as those infected by viruses, and kill them. For obvious reasons, NK cells are tightly regulated, but they generally respond to a kind of secreted **cytokine** called **interferon**, which is emitted by infected cells. Together, these various approaches to detecting and attacking invaders are called **innate immunity**. This system is responsible for many phenomena familiar to people, such as inflammation.

Parasites evolve as well, and a variety of organisms have developed methods for evading detection, or even surviving being consumed by phagocytes. Since the generation time of parasites tends to be much shorter than that of their multicellular hosts, the "arms race" of coevolution between parasites and hosts tends to tilt toward the parasites. Vertebrates have evolved a remarkable approach, called **adaptive immunity**, that allows long-generation-time multicellular organisms to keep up with their parasites.

The adaptive immune system takes longer to develop a response than the innate system, but it has **immune memory**, making it able to detect and respond to subsequent invasions of the same kind of organism much more quickly and effectively. Often the mechanisms of adaptive immunity are triggered by an innate immune response, strengthening it. Adaptive immunity can also recognize and respond to invaders that the innate system misses by developing sensitivity to highly specific molecular markers, called **antigens**.

The adaptive immune system can recognize far more different sorts of antigens than there are genes in the genome. This remarkable ability to recognize antigens from nearly any invader arises from **recombination**, a dynamic shuffling of parts of several genes in immune cells. Recombination is a regulated, somatic mutation process that produces an enormous variety of immune cell types, each specialized for recognizing invaders presenting a specific antigen. The cells of the adaptive immune system are called **lymphocytes**, and they make up about a third of the white blood cells (the rest are phagocytes from the innate immune system). Less than half of the lymphocytes are generally found in the blood; the rest are in a specialized lymphatic circulation system, consisting of the lymph nodes, lymphatic ducts, and the thymus gland.

There are two sorts of lymphocytes: **B-cells** and **T-cells**. B-cells produce proteins called **antibodies** that circulate in the plasma, called the **humoral immune response**. Antibodies, also called **immunoglobins**, are generated through the recombination process. Each recognizes and binds to one specific antigen. The very attachment of the antibody sometimes neutralizes an invader, and, if not, it triggers additional activity, such as activating the complement system or recruiting other immune cells. There are several classes of antibodies, each of which performs a specialized function. For example, the immunoglobulin A family (**IgA**) is found primarily in mucus, and is specialized for harsh environments such as the digestive tract.

B-cells arise in the bone marrow (hence their name), and the variable regions of their genomes undergo several steps of recombination as they mature, producing millions of different kinds of cells, each producing a specific antibody. During the maturation process, it is possible for a recombined protein to immediately recognize some molecule. Molecules recognized at this stage are self-antigens, so those B-cells are suppressed, ensuring that no antibodies to any of the body's own molecules are produced. Mature B-lymphocytes circulate through the blood and lymph systems, but do not start excreting antibodies until they are activated. Each mature, but inactive, B-cell expresses a little of its antibody in a form that is attached to the outside of its cell membrane. The first step to activation is the binding of that antibody to an antigen, although that alone is not enough to activate the cell. An interaction between the B-cell and a **helper T-cell** is required for the B-cell to become fully activated. The helper T-cell both activates the B-cell and transmits the information about the antigen to the T-cell system. Once activated, most B-cells become **plasma cells**, which produce large amounts of antibody but don't live very long. A few of the activated B-cells become **memory cells**, which don't produce much antibody but do live a long time. When memory cells encounter their antigen, they start producing more plasma cells immediately, meaning antibody production ramps up much faster on the second and subsequent exposures to an antigen.

T-cells produce the **cell-mediated immune response**, where antigens trigger the activity of special T-cells that are **cytotoxic**, meaning they can kill other cells. They are called T-cells because they undergo their maturation process in the thymus gland. Like B-cells, T-cells generate diversity through recombination, and exhibit memory, responding more quickly and intensely on second and subsequent infections. However, T-cells have more complex paths for maturation and activation, and use a somewhat different mechanism for recognizing invaders than B-cells. T-cells aren't responsive to whole proteins, but instead to peptide fragments of those proteins. Further, another cell has to first digest the foreign protein, attach a foreign peptide to a molecule called the **major histocompatibility complex** (MHC), and then present both molecules on its surface. This is what the B-cell is doing when it interacts with the helper T-cell; natural killers and other cells in the innate immune system also present peptides and MHC to T-cells.

The vertebrate immune system is a complex and sophisticated set of layered, interacting defenses. Yet despite its sophistication, vertebrates remain subject to a wide variety of invaders. The comparatively rapid evolution of parasitic organisms and the strong selective pressure that innovations in immune response place on them ensure that this will always be the case. On the other hand, the selective pressure that parasites put on hosts has driven the formation of this remarkably elaborate and generally quite effective system.

8.3.3 The Nervous System

As illustrated by the bacterial chemotaxis system described in section 4.4, mechanisms to sense the environment and modulate activity in light of sensations are very ancient adaptations. Early multicellular organisms could coordinate the activities of their cells through molecular signals. However, there are many sorts of potentially valuable communication and coordination that diffusing molecules cannot accomplish. For example, the use of muscles for motion requires alternating contractions of specific subsets of muscle cells, difficult to achieve with global molecular signals. The time it takes for such signals to traverse a large body also mitigates their usefulness in responding quickly. In multicellular animals, the benefit of being able to specifically coordinate activity among different types of cells, to deliver various messages to different cells simultaneously, and to do so quickly, gave rise to neurons, a cell type that is specialized for these functions. Neurons evolved relatively early in the history of multicellular organisms. Even in very simple animals like hydra, nerve cells rapidly deliver sensory signals that can cause dramatic changes in the organism.

In many organisms, a large portion of the cells in the body are part of the nervous system; for example, the nervous system of the adult nematode worm *C. elegans* includes about a third of its cells. In all of the Bilateria, sensory and motor neurons ramify extensively throughout the body. Axons from these neurons bundle together to form nerves.

Specific sense organs (such as the **retina** in the eye or the balance organ in the middle ear, called the **labyrinth**) consist of large numbers of specialized sensory neurons connected in particular patterns, with various types of supporting cells mixed in. These sensory cells contain specialized receptor proteins (many of them homologous to the G protein–coupled receptors that yeast evolved to find mates) that change state in response to light, sound, pressure, temperature, and so on. The sense organs that contain these cells are elaborated in ways that increase the sensitivity, range, and precision of the underlying sensor proteins.

The nervous system also controls most of the activity within an organism. Every muscle has an associated nerve that joins it at a **neuromuscular junction** that determines when the muscle will contract. Nerves also target the glands that secret hormones, influencing that communication system as well. Sensory and motor nerves converge on dense networks of interneurons, which are divided into two broad regions: the **spinal cord** (called the **ventral nerve cord** in arthropods) and the **brain** (the **cerebral ganglion** in arthropods). Interneurons connect sensations to actions. Simple, direct connections, sometimes involving only a single interneuron, are called **reflexes**.

Transformation of the rich sensory signals into effectively coordinated motor activity in pursuit of an organism's goals generally involves complex processing of the information in those signals. The nervous system is able to reliably identify salient signals, such as those that indicate the presence of food, mates, offspring, or various dangers, and to execute complex action patterns appropriate to the environment of the moment. The nervous system also plays a central role in homeostasis, directly regulating parameters such as heart and respiration rates, and organizing the overall behavior of the organism through drives such as hunger, thirst, and sexual desire.

Specialized formations of interneurons form brain regions that achieve these difficult tasks. For example, the **cerebellum** contains a variety of specialized neurons in a particular arrangement that facilitates the production of long and complex action sequences, such as walking. Another brain region, called the **visual cortex**, integrates signals from the retina. Particular populations of cells in the visual cortex respond preferentially to particular shapes, movements and even facial expressions through layered patterns of connectivity including both excitatory and inhibitory synapses. Another region of the brain, called the **limbic system**, is involved in the overall evaluation of the state of the organism, producing reward and fear signals as well as more complex emotions.

The details of the connections among neurons within and between brain regions play a large role in their functioning. For example, cells in the visual cortex respond strongly when sharp contrasts in the visual field are seen in a particular orientation. One set of cells responds to vertical lines, another to horizontal, and still others to each orientation in between. These cells are anatomically and physiologically indistinguishable from each other; what gives each its particular sensitivity are the specific other neurons that connect to it. The functional abilities of the nervous system are largely emergent; that is, they depend on the specific relationships among the constituent neurons and cannot be predicted from the functioning of the individual cells in isolation.

The interconnections among neurons involve extremely complex and specific contacts among large numbers of cells. Interneurons and motor neurons get inputs from the axons of a large number of other neurons. Incoming signals impinge on a neuron through its **dendrites**, which form a complex branching structure called a **dendritic tree**. The axon of one neuron doesn't quite touch the dendrite of the next; instead, the signal is transmitted by the release of molecules called **neurotransmitters** across a small gap called a **synapse**, and the effect of those neurotransmitters on receptors in the membrane of the dendrite. Neurons use dozens of different neurotransmitters to communicate. **Excitatory neurotransmitters** make the receiving cell more likely to fire an action potential, **inhibitory neurotransmitters** make the cell less likely to. Figure 8.4 shows a schematic representation of the process. Each interneuron

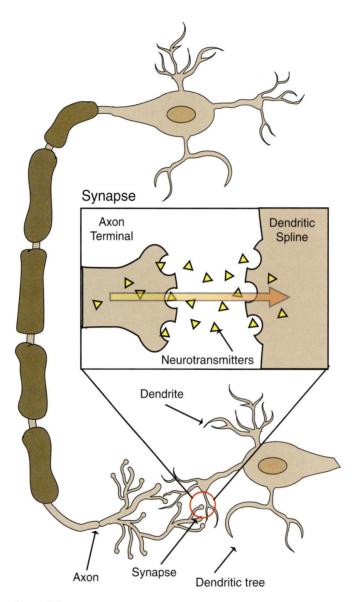

Figure 8.4

A schematic of transmission of information in the nervous system. The process begins when one neuron (at the top of the image) sends an action potential down its axon. The rate of transmission is increased by myelin sheaths that wrap the axon. The axon ends by making synapses with dendrites on receiving neurons. At the synapse, the axon terminal releases chemical neurotransmitters, which then interact with receptors on the dendritic spine, conveying a message to the receiving cell.

gets inputs from thousands of other neurons, and sends outputs to thousands more.

Phenotypic plasticity, the ability of an organism to change itself based on interactions with the environment, is perhaps most dramatic in the nervous system. **Learning** is the ability of an organism to change how it relates sensations to actions through experience in its environment. Functional changes in perception and behavior are created through structural changes in neurons, primarily involving the creation, loss, and changes in strength of synapses. The molecular basis for changes in neural connectivity and synaptic strength are increasingly well understood. One kind of change increases the strength of a synapse, called **long-term potentiation**, or **LTP**. Many kinds of LTP can take place in different sorts of neurons under different stimuli. One of the most basic occurs when a synapse using the amino acid glutamate as a neurotransmitter is exposed to an intensive burst of repeated action potentials. One of the receptors in the receiving (**postsynaptic**) cells that use this neurotransmitter is the **AMPA receptor**, which, when activated, allows calcium to flow through it into the cell. When a burst of repeated stimulation causes the calcium level within the cell to exceed a threshold, it activates **calcium-dependent kinases**, which have a variety of downstream effects. One effect is to cause the phosphorylation of the AMPA receptor itself, making it more sensitive, causing a temporary increase in the strength of the synapse. In the longer term, signal transduction pathways that begin with these kinases can result in transcription factor activation, ultimately producing new proteins that strengthen, or even completely remodel the synapse. Many other, more complex mechanisms alter synaptic strength based on the timing or sequencing of multiple inputs. The growth of new synapses or the decay of existing ones can be regulated by the activity of a set of neurons over an extended period of time.

Psychology (including studies of animal behavior) characterizes the function of the nervous system at a more abstract level, making use of theoretical constructs such as "perceptions," "goals," or even "trust." Recent progress in neuroscience has begun to link activity in specific regions of the nervous system to psychological constructs, both enriching the theoretical constructs and providing useful characterizations regarding the function of particular anatomical and physiological observations.[1]

---

1. A sampling of the accessible literature on this enormous topic includes Read Montague's *Why choose this book?* (Dutton, 2006); *Cognition, brain and consciousness: Introduction to cognitive neuroscience*, edited by Baars and Gage (Academic Press, 2007); and Byrne and Roberts' *From molecules to networks: An introduction to cellular and molecular neuroscience* (Academic Press, 2003).

8.4   Systems Biology

Though physiology has long taken an integrative approach that explains phenomena through complex relationships among components, much of molecular biology of the twentieth century focused exclusively on analysis of the components (biological molecules), in relative isolation from other phenomena. The value of the strictly reductionist approach is undeniable, but a movement to reemphasize the importance of context and interactions in understanding biological molecules has begun, generally under the rubric of **systems biology**.

The study of molecular interactions, such as those that underlie signal transduction pathways or molecular machines like the ribosome, have long been a staple of molecular biology. What is new in the systems biology approach is the emphasis on the molecular interaction network, rather than individual components. General properties of biological networks have been postulated, and evidence of their importance is growing. For example, the distribution of the number of interaction partners in many biological networks appears to be scale-free, a mathematical property that can be roughly translated as meaning that there are a few molecules with a large number of interaction partners and a far larger number that have only a few interaction partners. Assessments of connectivity are already playing a role in areas such as selecting targets for developing drugs and designs in genetic engineering (see sections 9.7 and 10.3). Related approaches, such as looking for particular topologies of connectivity that appear more frequently than would be expected by chance, have also demonstrated their value in identifying the likely function of previously uncharacterized proteins. Network effects can occur at all levels in biological systems, illuminating phenomena as diverse as metabolic control and the spread of epidemics through a population.

The current practice of systems biology also emphasizes a new sort of experimental methodology. Traditional molecular biology was strictly **hypothesis driven**. That is, each experiment was designed to test a specific hypothesis (e.g., whether a particular protein phosphorylates a particular amino acid on another particular protein in the presence of calcium). Alternatively, the systems biology approach has emphasized the analysis of complete, systemwide data, such as obtained by measuring the concentration of all RNA transcripts in a cell simultaneously, and the use of theoretical models that are tested against such datasets. Experimental methods from systems engineering, such as the analysis of function through systematic **perturbations**, are also part of this approach. Technological developments, described in section 10.1, have made many varieties of systemwide data available, and computational approaches have made predictive modeling practical in some biological areas. Reductionist, hypothesis-driven approaches and systems biology are not incompatible, and both are likely to make substantial further contributions to the human understanding of the process of life.

8.5 Suggested Readings

Anatomy and physiology are the subject of numerous textbooks. Perhaps the most widely used textbook of human anatomy and physiology is Tortora and Derrickson's *Principles of Anatomy and Physiology* (Wiley, 11th ed., 2005). Anatomy naturally lends itself to illustration, and there are many beautiful and compelling picture books in the field. Illustrated histology texts can be even more illuminating than those for gross anatomy; consider Ross and Pawlina's *Histology: A Text and Atlas with Correlated Cell and Molecular Biology* (Lippincott, Williams and Wilkins, 5th ed., 2006), which comes with an interactive CD-ROM.

Systems biology is so new that there are relatively few accessible texts. Although often more philosophical musing than explanation, Denis Noble's *The Music of Life* (Oxford Press, 2006) clearly expresses the systems biology perspective. For those with some background in physics, Uri Alon's *An Introduction to Systems Biology: Design Principles of Biological Circuits* (Chapman and Hall, 2006) is a compelling read.

9   Disease and Its Treatment

All living things can get sick. Even bacteria can be infected by viruses that cause a loss of reproductive fitness, or even death. Organisms face a lifetime of damaging stresses, including injury, infection, malnutrition, and environmental toxins. To varying degrees organisms are able to adapt, and often recover from the damage inflicted. In addition to the immune response aimed at neutralizing invaders, organisms have a variety of mechanisms to stave off or repair the damaging effects of these stresses. Our endogenous responses can also be aided by external interventions, that is, by medicine.

Advances in medicine, particularly over the last century, have greatly reduced the burden of injury and disease, extending and improving the quality of human life. Molecular biology has made significant contributions to the understanding of the mechanisms of disease and healing, as well as to the creation of safe and effective therapeutic interventions. Medicine is a vast topic with an enormous literature. The discussion in this chapter is necessarily limited to a few broad principles and illustrative examples, particularly emphasizing the role of molecular phenomena. In this chapter, the focus is on human beings. Veterinary medicine is largely, but not entirely, similar to human medicine, and each species has its own particular strengths and weaknesses with respect to health.

It is important to note that, although molecular biology has had an important role in the development of new pharmaceuticals for some time, the introduction of molecular concepts into the clinical care of patients is relatively new and still modest. However, the rapid pace of development of molecular understanding and biotechnology has led to the widespread prediction of radical changes in the future of medicine. Those speculative topics are discussed in section 11.1; the following sections focus on medicine as it is practiced today.

9.1 General Principles of Pathology and Therapeutics

The broadest definition of **disease** is any impairment of normal biological function, including injuries, malnutrition, poisoning, infections, genetic abnormalities, metabolic abnormalities, electrolyte abnormalities, malignancy, and sometimes is defined to include the effects of aging. The medical community characterizes disease based on its **etiology** (cause), its signs (clinically observable phenomena) and symptoms (sensations felt by the patient), its **pathogenesis** (the underlying mechanisms that cause the signs and symptoms), its natural history (what happens during the course of the disease), and its treatment.

Specific organ systems respond to pathologic stimuli in different ways, but the molecular, cellular, and tissue responses have important commonalities. Ultimately, all disease arises from abnormalities in cell structure and function. Cells actively respond to stresses, and adjust their functioning to maintain homeostasis. When cells cannot adapt to stresses, they develop characteristic cellular injuries. These injuries lead to further alterations, some of which may be reversible, but also can include permanent changes or the death of the cell. The general mechanisms described in this section underlie the specific mechanisms of the diseases discussed in later sections.

Since before the beginning of recorded history, people have tried to treat the sick and the injured. Ancient approaches included surgery, nonsurgical procedures, and the use of medicines. Contemporary approaches also use these approaches, although with far greater success. That success derives in large part from the scientific understanding of the structures and functions of the human body, and the growing molecular understanding is likely to further expand the repertoire of successful medical interventions.

9.1.1 Cellular Stress and Responses to It

Many different sorts of stresses can affect a human cell. Perhaps foremost among them is lack of oxygen, called **hypoxia**, which has immediate effects on the metabolism of a cell, and can kill it within minutes. Chemical agents of many kinds can harm cells. Poisons inflict damage through a wide variety of mechanisms, and even innocuous substances such as salt or water can become stressors in sufficient concentrations. Disease-causing infectious agents, called **pathogens**, including bacteria, viruses, fungi, or even parasitic animals like tapeworms, can also inflict many kinds of damage. Imbalances resulting from either too little or too much of various nutrients in the diet sadly remain a major cause of cell injury. Physical forces, including trauma, temperature extremes, rapid changes in pressure, radiation, and electric shock can all damage cells. Endogenous sources of cellular stress are also important, including genetic defects, inappropriate immune responses and aging.

Cells adapt to these stresses in four basic ways: **Hypertrophy** is an increase in the size of cells, generally leading to an increase in the size of the organ containing them. Hypertrophy can be an appropriate response to stress (such as when a muscle grows after exercise), or can be disease related (such as when heart muscle grows in response to high blood pressure). Hypertrophy can be triggered by molecular signals. For example, brief exposure to the hormone adrenaline causes increases in heart rate and in the power of its contractions. Extended exposure can cause cardiac hypertrophy, and ultimately cardiac injury as the heart cells reach the limit beyond which increases in cell size can create increases in power. This response is an example of how an initially adaptive response can ultimately lead to cell injury if a stress persists for a long time.

Hypertrophy tends to occur in cell types, like cardiomyocytes, that cannot divide. A typical response to the same sort of stress in a dividing cell population is **hyperplasia**, or increases in the numbers of cells. Hyperplasia can also be normal (such as regrowth as part of healing a wound) or part of a disease process (such as the growth of a wart in response to a viral infection). Hyperplasia occurs in response to molecular signals of hormones or growth factors. The difference between hyperplasia and cancer is that hyperplasia ceases when the molecular stimulus is removed, whereas cancer cells multiply without regard to external molecular signals (see section 9.6).

The third cellular adaptation to stress is **atrophy**, shrinkage of cell size. Diminished blood supply, loss of hormonal stimulation, malnutrition, disuse, or even just aging can force cells to adapt to reduced resources by reducing their metabolic demands. In extreme circumstances, cells will undergo **autophagy**, where they digest their own components to obtain enough nutrients to survive.

The final mechanism of cellular adaptation, called **metaplasia**, results in a change in the type of cells in a tissue. Metaplasia replaces cell types that are sensitive to a particular stress with cell types that are less sensitive. For example, the upper respiratory tract is normally lined with ciliated columnar epithelium. In chronic cigarette smokers, these cells can be replaced by much tougher stratified squamous epithelium. These squamous cells are better able to survive chronic exposure to heat and smoke. This transformation comes at a cost, however; for example, the loss of cilia means that foreign particles cannot be effectively expelled, compounding the ultimate damage from smoking. Metaplastic changes are thought to result from changes to the stem cells, rather than transformations of differentiated cells, although the molecular mechanisms are still obscure.

9.1.2 Cell Injury

Although the initial adaptations to cellular stress described here can be effective for a time, continued stress leads to cell injury. The mechanisms of cell injury are best understood in molecular terms. Understandably, one of the most severe injuries a cell can suffer is the loss of ATP. Hypoxia, damage to mitochondria, loss of access to nutrients, and the actions of some poisons (e.g., cyanide) can cause a dramatic fall in ATP. Depletion of ATP to below about 10% of normal has devastating effects on a cell. One of the first molecular systems to fail without adequate ATP supplies are the pumps that maintain homeostatic concentrations of ions, leading to increases in intracellular sodium and calcium levels, and an influx of water. The resulting swelling of cells is an early indication of injury.

Cell injury can also be caused by other factors. Damage to any of the membranes within a cell, especially in the mitochondria, is an important mechanism of cell injury. Many toxins, such as that produced by the venom of some poisonous spiders, function by attacking membranes. Accumulation of damaged proteins or DNA cause cell injury, such as that seen in the brain degeneration associated with Alzheimer's disease. Influx of too much calcium can also cause reactions that damage the cell, as happens, for example, in mercury poisoning.

One of the primary molecular mechanisms of cell injury is the accumulation of **reactive oxygen species**, or **ROS**. Reactive oxygen species, sometimes also called free radicals, are unbound compounds with a single valence electron, and are hence very reactive. Such molecules are an important part of normal function, including as part of the electron transport chain used to produce ATP. However, they can damage many biological molecules (proteins, lipids, and especially nucleic acids) if their concentration is too high, or if they escape from their proper cellular compartments. A variety of antioxidants, particularly the protein superoxide dismutase, scavenge for inappropriate ROS, enzymatically rendering them harmless.[1] External forces, such as ultraviolet light from the sun, ionizing radiation, or certain poisons (e.g., carbon tetrachloride) can

---

1. Unfortunately, there is no evidence that high doses of antioxidant molecules in the diet have any effect on the activity of this system.

increase the concentrations of ROS beyond the ability of the antioxidant systems to cope, creating **oxidative stress**, and ultimately causing cell injury. ROS react with biological molecules, oxidizing them. Oxidized proteins, lipids, and DNA cannot function normally, and they can interfere with many other aspects of normal function.

Internal events can also cause increases in ROS. For example, the innate immune system produces free radicals both to kill invading bacteria and as a signal to recruit fibroblasts and begin the process of wound healing. The oxidative stress due to inflammation can injure other cells.

Ischemia, or loss of blood flow to a tissue, causes stress and possibly injury due to hypoxia and loss of nutrients. Ischemia is the most common cause of cell injury in clinical medicine. The injury or death of cells due to ischemia is called **infarction,** and the injured area is called an infarct. Oxidative stress also plays an important role in ischemic stress. Reperfusion (the return of blood to an ischemic area) can result in oxidative stress that perversely is even more damaging that the original loss of blood. Understanding of this molecular mechanism of cell injury has turned out to be particularly important in the care of heart attacks and strokes.

9.1.3 Cell Death

If the causes of cell injury are not reversed, the cells die. There are two ways that cells can die, and they have quite different consequences for the organism. **Necrosis** is the messy way to go. The swelling that begins with cell injury continues, and cellular organelles either swell (e.g., mitochondria and endoplasmic reticulum) or disappear (ribosomes). The cell membrane loses its integrity, and the contents of the cell leak into the interstitial fluid. Rupturing of the lysosomes releases digestive enzymes that degrade the remains of the cell. Cytoplasmic proteins associated with particular cell types are released in this process and end up in the blood, where they can be assayed to determine the source and extent of cell death. The innate immune system generally responds to necrosis, sending white blood cells into the area and causing inflammation. Necrosis is the result of a failure of cellular metabolism, and only occurs as the result of a disease process.

The other sort of cell death, called apoptosis, is under the control of the cell, and is sometimes called programmed cell death or cell suicide. It's a quiet death, in which cells wither away rather than burst, encouraging nearby phagocytes to consume bits of the dying cell, recycling its components. No inflammation or other immune response occurs. It can be difficult to detect even substantial amounts of apoptotic activity clinically.

Apoptosis is a part of normal development (as discussed in section 7.2.2), but also can occur in response to cell injury. The activation of apoptosis involves the release of enzymes called **caspases**. Caspases chew through membranes, degrade the cytoskeleton, and activate other proteins called endonucleases that chew up DNA. Once the caspase cascade begins, the cell is fated to die quickly.

All cells carry around the extremely dangerous caspase proteins at all times. For safety they are synthesized as part of a large polyprotein, rendering them inactive. Two mechanisms can set them loose, one that is sensitive to the ATP production within a cell, and one that listens to signals from outside of the cell. The internal system depends on a mitochondria protein called cytochrome c, which normally functions as part of the electron transport chain. When cytochrome c is released into the cytoplasm, it frees one of the caspases from its polyprotein, which in turn activates the other caspases and begins the apoptotic cascade. A variety of mechanisms can trigger the release of cytochrome c from the mitochondria. Most basically, damage to the mitochondrial membrane will allow it to leak out. However, a variety of internal cellular sensors can also trigger the release of cytochrome c when DNA is damaged irreparably, when too many misfolded proteins accumulate, when required growth factors are not present (in hormone and cytokine sensitive cells), and when calcium levels get too high.

Apoptosis can also be initiated through a signal transduction pathway, triggered by proteins called the death receptors. There are several kinds of death receptors, including highly studied proteins such as tumor necrosis factor (TNF) and FAS. The signal processing for apoptosis is some of the most complex in all of biology. It balances death receptor signals (called proapoptotic) with signals received from growth factor receptors (called antiapoptotic) in a complex and incompletely understood pathway that also takes into account other aspects of the state of the cell. External triggers of apoptosis (other than during normal development) are generally T-lymphocytes that have detected a cell is infected with a virus or other pathogen.

9.1.4 General Principles of Therapeutics

Medical procedures are what clinicians (doctors and allied professionals) do to alleviate or cure disease. There are two broad classes of medical procedures: diagnostic (intended to determine the problems afflicting a patient) and therapeutic (to cure or alleviate those problems). Although diagnostic procedures are a vital prerequisite to most therapy, and they increasingly make use of molecular biotechnology, they form too large and diverse a set of topics to even summarize here.[2] Contemporary therapy includes surgery (which involves cutting into the body), pharmacotherapy (administration of drugs), and medical procedures (such as setting broken bones, blood dialysis, therapeutic irradiation, or psychotherapy).

The best approach to any disease is prevention. **Immunization** involves sensitizing the adaptive immune system to a pathogen through the use antigens that do not cause disease. Vaccinations use portions of the pathogen, dead pathogens, or toxins produced by a pathogen that have been somehow rendered harmless. Once sensitized, the adaptive immune system is able to mount a much more rapid and vigorous response to the pathogen, usually preventing any occurrence of the disease. Immunization is the only practical treatment for many viral infections. **Public health** measures, such as ensuring general access to clean water or screening for hidden but treatable conditions like high blood pressure, also have a major role in disease prevention.

Medical interventions are not able to cure all diseases. Sometimes patients are treated symptomatically, meaning that interventions are intended to reduce their symptoms (such as pain relief or fever reduction) rather than cure the disease. This is particularly common in diseases that can be expected to resolve on their own, such as many viral respiratory infections. Supportive therapy (such as ventilator assistance for breathing) helps patients live with their symptoms, usually to provide time for other interventions to work. Palliative therapy concentrates on relieving suffering rather than effecting a cure, and is offered to severely ill patients.

The most basic therapeutic task is to ensure the continuance of life, called life support. Patients with immediately life-threatening conditions must be **resuscitated**, which aims to ensure that the respiratory and cardiovascular systems are able to function well enough to prevent immediate death. Many factors can immediately threaten life, and an even larger number of possible interventions address them, ranging from first aid to transplant surgery. Interventions to resuscitate critically ill patients can involve clinical use of the body's signaling molecules, such as adrenaline, which can improve cardiac output and blood pressure.

Curative interventions are meant to address the cause of the disease, or repair the damage done through injury, rather than merely alleviate symptoms. Surgery can remove or repair abnormal tissue, remove foreign bodies that are causing abnormalities, implant mechanical aids (such as pacemakers to

2. For an entertaining discussion of the ongoing revolution in medical diagnosis from a credible prognosticator, see Andy Kessler's *The end of medicine: How Silicon Valley (and naked mice) will reboot your doctor* (HarperCollins, 2006).

regulate heartbeat, stents to reopen blocked passages, or artificial replacement joints), or transplant healthy tissue to repair a defect. Surgical interventions require drugs and other molecular techniques to succeed but are inherently macroscopic.

Many of the successes of modern medicine, particularly in treating infectious disease, have come through the discovery or invention of pharmaceutical small molecules, that is, drugs. Currently 3,898 drugs are approved for sale in the United States.[3] While attempting to provide a brief overview of disease and its treatment generally, this chapter focuses particularly on the molecular, and therefore on pharmaceuticals.

Drugs have been particularly effective at curing infectious disease. Many of the scourges that killed so many people through history have been addressed by drugs developed in the last hundred years or so. The antibiotics that treat bacterial infections and immunizations that prevent viral infections have saved countless lives. These drugs are discussed in section 9.4.

In much of the developed world, infectious disease is no longer the primary threat to health. Environmental and genetic disorders, such as cancer, diabetes, and cardiovascular disease take a much larger toll in those parts of the world. The treatment of these diseases is complicated, but many of the interventions available are based on understanding of the relevant molecular biology.

9.2 Healing

One of the most remarkable abilities of organisms is to heal, to recover from injury. **Lesions**, or areas of damaged tissue, can arise through trauma, infection, or injury sustained as a result of hypoxia or ischemia, among other causes. When cells are damaged, the immediate response involves coordination of the circulatory system and the immune system. There are three stages to repair: inflammation, which seals off the damage and sends out signals to begin repairs; proliferation, when rapid cell growth begins to replace the lost cells; and remodeling, when the replaced cells organize into functioning tissue and initial scaffolding tissue dies off through apoptosis. The processes are continuous, and the stage boundaries are somewhat arbitrary.

The first stage of wound healing involves sealing off any damaged blood vessels. Breach of a vessel is detected when platelets (clot-related blood cells)

---

3. This is the number of National Drug Codes, a universal product identifier for human drugs from the US Food and Drug Administration, as of August 2007. This is an overestimate of the number of different compounds in use, as different dosages and formulations (e.g., time released) of a particular compound count as different drugs. There are more than 1,000 different drug compounds, although a small proportion of those make up the vast majority of uses.

come into contact with the extracellular matrix protein collagen. These platelets become activated, expressing membrane glycoproteins that stick to each other and to the site of the damage. Ruptured membranes release prostaglandins, which are lipid compounds that trigger spasm of the smooth muscle lining the vessels, restricting blood loss. Activated platelets also release signal peptides that recruit other platelets and trigger the **coagulation cascade**. This cascade is highly conserved among vertebrates. More than a dozen proteins interact in the cascade, integrating information both from contacts with collagen and through signaling from damaged tissues. Many of the proteins in the clotting cascade are **serine proteases**, which means that they are enzymes that cut proteins (proteases) using the action of serine residues in the active site. The cleaved proteins become functional, activating the next step in the pathway, ultimately producing a burst of a protein called thrombin. Thrombin keeps the cascade going, recruits additional platelets to the site, converts fibrinogen into fibrin (the main protein that forms a clot), and activates other proteins that cross-link the fibrin and the platelets into a tough, impermeable polymer.

As the formation of a blood clot seals off damaged blood vessels, a variety of signaling molecules trigger the initiation of inflammation. In the inflammatory phase, the prostaglandin-triggered constriction of vessels is replaced by peptide signals such as bradykinin and histamine, which cause the local blood vessels to dilate and become permeable, and the smooth muscle lining the vessels to relax. Plasma and white blood cells start leaking out of the nearby capillaries. The complement system is activated, increasing the influx of fluids. White blood cells are attracted to the site of the damage by lipid signals, called leukotrienes. Phagocytes arrive, and consume both damaged tissue and any bacteria that may have invaded the site. Another kind of white blood cell, the granulocyte, arrives later, and uses reactive oxygen species to kill bacteria and break down damaged tissue.

Inflammation is necessary to fight infection and induce the proliferative stage that comes next, but extended inflammation can lead to additional tissue damage. Inflammation continues as long as there are foreign bodies in the wound, which is one reason it is important to wash wounds thoroughly. Anti-inflammatory drugs can also be used once threat of infection is over.

Finally, several days after the injury, white blood cells called macrophages become the dominant immune cells. Like phagocytes, they consume bacteria and damaged tissue, but they also secrete growth factors and cytokines that pave the way for the proliferative stage.

The proliferative stage reconstitutes the extracellular matrix and the formation of granulation tissue. Granulation tissue fills the void left by the wound, and contains new blood vessels, fibroblasts, inflammatory cells, and the precursors to extracellular matrix. Fibroblasts produce the proteins fibronectin, laminin, and integrin, which are used to build a temporary extracellular matrix. The fibroblasts initially follow the fibrin in the clot, proliferating and leaving a trail of ground substance and later collagen as they go.

The next wave of cellular regeneration follows these matrix proteins. The matrix is labeled with growth factors, such as angiogenesis (blood vessel growth) signals, that recruit new epithelial cells to the area and organize the cells into appropriate structures. Once the matrix is adequately reconstituted and the local blood supply is reestablished, normal cells appropriate for the tissue begin to migrate into the area.

In the remodeling phase, the temporary structures are digested and replaced with more normal tissue. Contraction, mainly based on atrophy and transformation of the fibroblasts, reduces the size of the wound. As the tissue contracts, blood vessels that are no longer needed are removed by apoptosis.

In some tissues, full tissue regeneration is possible after injury. In others, normal tissue cannot be fully reconstituted, and the collagen matrix remains as scar tissue. The molecular triggers that could fully recapitulate developmental programs in order to regrow damaged tissue are the subject of intense study, but are not currently well understood.

9.3 Genetics and Disease

The use of genetics to understand human disease began almost immediately after the rediscovery of Mendel's work in the early twentieth century. Early work traced the occurrence of hemophilia (a failure to form blood clots) in the British Royal family, which followed the Mendelian inheritance pattern for a gene present on the X (female) chromosome (see section 2.1). The first human linkage analysis was published by J.B.S. Haldane and Julia Bell in 1937, showing that hemophilia was linked with red-green colorblindness. Since then, thousands of human phenotypes, mostly disease-related, have been shown to have Mendelian inheritance.

Most of these inherited diseases are relatively rare congenital (i.e., present at birth) metabolic defects. A typical example of this sort of disease is phenylketonuria, better known as PKU. PKU is caused by a defect in the enzyme phenylalanine hydroxylase (PAH), which converts one amino acid, phenylalanine, into another, tyrosine. Currently, more than 400 PAH mutations have been observed in PKU, some of which lead to more severe forms of the disease than others. Without this enzyme, phenylalanine concentrations grow to the point where they can cause damage. Most important, a high concentration of phenylalanine in the blood interferes with the transport of other amino acids into the brain, causing abnormal brain development. If the disease is diagnosed early enough, dietary restriction of phenylalanine (and supplementation with tyrosine) can avoid the worst consequences of the disease. All newborns in the United States (and in many other countries) have a drop of blood drawn shortly after birth to test for PKU.

Until recently, medical genetics was a minor part of pediatrics, since most Mendelian traits associated with disease cause rare congenital illnesses. However, it has long been known that many other diseases have heritable components. Diabetes, heart disease, and some kinds of cancers tend to run in families. As explained in section 2.1, the fact these diseases don't show Mendelian inheritance patterns suggests that multiple genes are involved, and the disease involves interactions among them. It is much more difficult to identify the particular combination of alleles that might play some role (perhaps only in combination with particular environmental conditions) in these diseases.

The ability of new technologies to identify the hundreds of thousands of SNPs in the genome of a typical person (see section 10.1.5), combined with computational analysis of the patterns of inheritance of all of those polymorphisms, is beginning to produce information about which genes and interactions play a role in these complex phenotypes. Even once the contributing genes are identified, a substantial amount of research into the mechanisms by which these genes influence the development of disease, and into potential treatments, must be done before clinical treatment can be improved. One of the first examples of the discovery of a common gene variant with clinical consequences is the discovery that a particular allele of the gene TCF7L2 contributes to risk of type 2 diabetes. Interestingly, people with the high-risk allele do not respond as well as others to Metformin, the primary drug used to treat prediabetes, but do respond better than others to lifestyle changes. This means that environment is more important to the people who have this genetic variant, an interesting twist on most people's intuitions about the role of genes in disease.

One reason that a disease may appear to have a complex inheritance pattern is that the "disease" is really several different diseases that all manifest themselves in similar ways. Cancer is like that: It has many different causes. A few cancers are caused by single gene defects. For example, the well-publicized "breast cancer genes," BRCA1 and BRCA2, are inherited in a Mendelian pattern, and having two defective alleles does strongly predispose a woman to develop breast cancer. However, only a tiny proportion of breast cancers are caused by defects in BRCA genes. Cancer is discussed further in section 9.6.

9.4   Infectious Disease and Antibiotics

Many diseases are caused by specific infectious organisms, called pathogens. Treatment of these diseases involves killing or preventing the growth of the pathogen. Pathogens include viruses, bacteria, protists, and even some metazoan parasites, like tapeworms. Many of these diseases can be addressed by immunization. Those that cannot are often well treated by antibiotics. Literally, the word antibiotic suggests any substance used to kill other organisms, although clinicians use the term to refer only to antibacterial agents.

Many bacterial infections are localized and can be effectively treated with antibiotics. However, a bacterial infection of the blood, called **septicemia**, is particularly serious, since it can lead to generalized infection throughout the body. An overwhelming blood infection is called **sepsis**, which, if not controlled immediately, can lead to a dramatic loss of blood pressure called **septic shock**, and ultimately to multiple organ failure and death.

A wide variety of antibiotics are available, suited to most kinds of bacterial infections. Antibiotics attack bacteria using a variety of mechanisms, primarily either attacking the cell wall or inhibiting the bacterial pathways for synthesis of protein, RNA, or DNA. Significant differences between bacterial and human metabolism make antibiotics among the safest available drugs. Antibiotics differ in their effectiveness for specific bacteria; effectiveness also depends on the site of the infection within the body.

These drugs exert a severe selective pressure on the bacterial populations, resulting in the evolution of mechanisms of resistance. Indiscriminate use of antibiotics, and particularly the discontinuation of antibiotic therapy before all of the targeted bacteria are killed, can result in the creation of drug-resistant strains. Often drug resistance genes are encoded in plasmids, which are shared among bacteria (see section 4.3), worsening the problem.

There are comparatively few effective antiviral agents. One of the few effective antiviral drugs is aciclovir, which targets the family of viruses that cause herpes sores and shingles. Aciclovir has an interesting mechanism of action. It is processed by infected cells to form a nucleotide analog that competes with guanine for incorporation in replicating virus; when incorporated into the DNA it terminates prematurely, preventing replication. Aciclovir is converted into its active form only by one of the viruses' own proteins, ensuring that DNA replication is affected only in infected cells. Aciclovir doesn't kill the virus, but it does interfere with its ability to reproduce, and reduces the duration and severity of its symptoms.

The most significant viral disease, acquired immunodeficiency syndrome (AIDS) is caused by human immunodeficiency virus (HIV). HIV is a very unusual virus. It is a member of a small class of organisms now known as retroviruses. The retroviral genome is stored as RNA, which must be reverse-transcribed to produce DNA that the cell then uses to direct the production of more virus. The HIV genome encodes a protein, called reverse transcriptase, that manages this contravention of the central dogma.

HIV infection is treated with two types of drugs. The first class of anti-HIV drugs interferes with the reverse-transcription process, which slows the replication of the virus. The second class of drugs takes advantage of the fact that the viral proteins are synthesized as a large polyprotein, and then cleaved into active constituents by a protein called the HIV protease. Drugs called protease inhibitors aim to prevent the cleavage of the polyprotein and therefore the formation of active viral proteins. While these drugs can substantially delay the onset of AIDS, there is as yet no immunization or cure.

Aside from these few specific classes of antiviral drugs, there are two less effective, but more general approaches. One is the administration of interferon. Interferon is a naturally occurring cytokine produced by the immune system that regulates various immune responses. Human interferon protein can be produced by genetic engineering (see section 10.3), and its use is of demonstrated value in treating certain viral infections, particularly hepatitis C. The other approach to the treatment of these infections is to inject antibodies targeted at viral proteins, called passive immunization. Although these antibodies do not last very long in the body and are not replaced by the immune system, they can be effective against tetanus, diphtheria, and botulism.

Infections by protists, fungi, and helminthes (worms) are much less common than bacterial or viral infections, but they can be hard to treat. The fundamental similarity of the biochemistry among eukaryotes makes pharmaceutical therapy more challenging. Many drugs used to treat these infections have significant side effects. Malaria, one of the most devastating and difficult to treat diseases globally, is caused by a protist, and though drug therapy is important in alleviating the symptoms and preventing progression, it is not currently possible to cure malaria.

9.5 Cardiovascular Disease

Cardiovascular disease is the leading cause of death in the developed world. There are many different cardiovascular diseases, but the bulk of deaths are caused by ischemic disease (heart attacks), heart failure (not producing adequate circulation), and arrhythmias (disorders of heart rhythm).

Each of these types of cardiovascular disease can lead to sudden death. There is a very short period of time between cardiac arrest (when the heart stops functioning) and irreversible damage; brain hypoxia causes injury within about four minutes, and irreversible injury within seven minutes. Cardiopulmonary resuscitation, or CPR, is a procedure that can maintain some circulation and oxygenation, delaying irreversible injury until more advanced life support can begin. Since emergency response time in the United States averages about seven minutes, CPR can be extremely important. More than 1,000 people a day die of sudden cardiac arrest before they get to an emergency room. Fewer than half of the people who have a serious heart problem outside of a hospital get CPR before emergency personnel arrive. Yet CPR instruction for laypeople is widely available and as little as half an hour of training every two years can provide the necessary skills.[4]

Ischemic heart disease is caused by inadequate blood supply to the heart. The heart is supplied with blood by the coronary circulation system. Most cardiac ischemia is caused by mechanical blockage of the coronary arteries, called coronary artery disease, or CAD. CAD remains the largest single cause of death in the United States, even though the mortality rate has fallen by more than a third in the last 20 years.

The pathogenesis of CAD involves progressive **atherosclerosis** of the coronary arteries. Atherosclerosis[5] is a narrowing of the arteries due to the buildup of plaques that are made of lipids, macrophages, and smooth muscle cells. A combination of factors, including mechanical stresses (e.g., from high blood pressure), biochemical abnormalities of the blood (e.g., from diabetes or elevated levels of circulating lipids), and immunological factors (e.g., ROS or inflammation) can create an initial injury to the arteries. This injury causes the invasion of white blood cells and the accumulation of **lipoproteins** (packages of proteins and lipids circulating in the blood; see following text). Invading macrophages ingest the lipid, oxidizing it with ROS, which gives them a characteristic appearance, evocatively described as a foam cell. These cells produce inflammatory cytokines, including members of the TGF-β superfamily mentioned in section 7.3, and induce the production of C-reactive protein, a diagnostic marker for inflammation. These molecular signals trigger the growth of smooth muscle cells into the developing plaque, and recruit additional monocytes and macrophages from the bloodstream, reinforcing the

4. For more information about CPR and how to get training, see http://www.nlm.nih.gov/medlineplus/cpr.html.

5. *Arterio*sclerosis is a general term describing any hardening (and loss of elasticity) of medium or large arteries; *athero*sclerosis (pronounced ath-uh-ro-skluh-RO-siss) is a kind of arteriosclerosis caused by the deposition of plaque. "Athera" is the Greek word for porridge, and the plaque were said to look like lumps of porridge. Pathologists often use Greek food words to describe the appearance of tissues (e.g., **caseous**, "cheesy" necrosis).

process. The smooth muscle cells produce increasing amounts of collagen, and the cells and collagen can grow into a fibrous cap over the core deposit of lipids and immune cells, together called an advanced plaque.

When plaques grow to the extent that they substantially reduce the cross-sectional area of the artery, normal blood flow requires maximum dilation of the vessels. In this situation, it is not possible to increase the oxygen supply to the heart and increased demand, for example during exercise, results in ischemia and chest pain (known as **angina**).

Two mechanisms can produce an internal blood clot (called a thrombus) from these plaques. The first involves disturbance of the surface of the fibrous cap, exposing its collagen to the circulating blood. As described earlier, blood clotting is triggered by the exposure of platelets to collagen. The second process occurs with a rupture or tear in the fibrous cap that lets blood reach the core of the plaque. The combination of collagen and cytokines secreted by the macrophages produce a large clot within the plaque, expanding its volume, and further opening the breach. Should clots arising through either process break loose, they can block already narrowed arteries and cause dramatic cardiac ischemia. The death of cells as a result of this ischemia is called a myocardial infarction (MI), better known as a heart attack.

The goal of treatment of MI is to reestablish blood flow to the heart as quickly as possible. If initiated quickly, surgical intervention to physically reopen blocked vessels is the most effective method. However, if immediate surgery isn't feasible, then drugs that dissolve clots are almost as effective. In either case, the time to treatment is a critical factor in effectiveness. Some of the drugs given to MI patients are very old: for example, aspirin blocks the activation of thromboxane, a lipid that platelets need to aggregate. Others are new: For example, recombinant tissue plasminogen activator (r-TPA) is a human protein produced industrially through genetic engineering (see section 10.3). Plasminogen is a protein that cleaves fibrin, which is the primary protein that holds clots together; r-TPA activates plasminogen already circulating in the blood, breaking up the clot. Even when surgical treatment is practical, drugs to block certain receptors on platelets have been shown to improve outcomes. One of these drugs, abciximab, is a commercially produced antibody.

Excess lipid concentration in the bloodstream, called hyperlipidemia, is an important contributing factor to plaque formation. Since lipids are insoluble in water, they are transported in a capsule of protein; together the lipid and the protein are called a **lipoprotein**. Many sorts of lipoproteins circulate in the blood. Lipoprotein particles that contain relatively large amounts of lipid (and less protein) are called low-density lipoproteins (LDLs). They transport particular lipids from the liver to the other cells in the body. Lipoprotein particles that have relatively more protein (and less lipid), called high-density lipoproteins, transport lipids back to the liver. In popular parlance, LDL is "bad cholesterol" and HDL is "good cholesterol," although actually both are more complex molecular assemblies that include proteins and sometimes lipids other than cholesterol.

Lipid metabolism is complex, and incompletely understood. **Steroids** are a broad class of lipids that contain four carbon rings. Steroids are widely used throughout the eukaryotes, including in plants, fungi, and vertebrates. Many signaling hormones are steroids, including the sex hormones estrogen and testosterone, the corticosteroids that regulate metabolism and electrolyte levels (and are used as anti-inflammatory drugs), and the anabolic steroids that signal to increase bone and muscle synthesis. Cholesterol is a sterol (a steroid with alcohol group (-OH) attached) that is synthesized in the liver. It is used throughout the body as a precursor for the synthesis of many other vital lipids, including the phospholipids that make up cell membranes (see section 4.1), vitamin D, and many of the steroid hormones. Cholesterol is a vital molecule that plays many important roles in the body, but an excess of it can be deadly.

Apolipoproteins are the proteins that package lipids for transport. LDL particles include a protein called apolipoprotein B, which is one of the largest proteins in the human body (with more than 4,500 amino acids). Defective apolipoprotein B genes cause severe hyperlipidemias, and some less dramatic polymorphisms have been suggested to affect risk of cardiovascular disease. Research into the structure and function of these proteins and their polymorphisms is intense. For example, some evidence suggests that large HDL particle size is associated with very long lifespans[6].

Pharmaceutical interventions to lower the blood level of cholesterol-containing LDLs are among the most prescribed drugs in the United States. The main drugs used to lower circulating LDL levels, called statins, are inhibitors of an enzyme called 3-hydroxy-3-methylglutaryl-coenzyme-A reductase. That enzyme catalyzes the rate-limiting step in the synthesis of cholesterol in the liver. Other drugs can also play a role in reducing or reversing atherosclerosis. For example, the smooth muscles are induced to constrict (raising blood pressure) by a peptide called angiotensin; a class of drugs called angiotensin-conversion enzyme (ACE) inhibitors block this process. ACE inhibitors both

6. See, for example, Arai and Hirose, Aging and HDL metabolism in elderly people more than 100 years old, *Journal of Atherosclerosis and Thrombosis* 11; 5 (2004), 246–252, or Barzilai et al., Unique lipoprotein phenotype and genotype associated with exceptional longevity, *Journal of the American Medical Association*, 290; 15 (2003, Oct.), 2030–2040.

reduce blood pressure and appear to have an independent effect in extending the life of patients with atherosclerosis.

9.6   Cancer

Cancer, the uncontrolled and invasive growth of mutated somatic cells, is the second most frequent cause of death in the developed world. Nearly a third of all people in the developed world will develop cancer at some time during their lives. Although the disease is often fatal, many advances in treating it have been made recently, and molecular understanding of the mechanisms involved is growing rapidly.

Cancer marks a fundamental loss of control in the multicellular body, where somatic cells multiply at the expense of the organism as a whole. There are many mechanisms that multicellular organisms use to prevent mutated somatic cells from causing damage to the body; cancer represents the failure of all of these many layers of control. These layers include both internal mechanisms within each cell and external immune system surveillance and response to damaged cells.

Cancerous cells initially grow in a specific location, forming a nonfunctional tissue mass called a **tumor**. Eventually, the cancer cells invade adjacent tissues, and often migrate through the blood or lymphatic circulation to other areas, forming **metastases**. Tumors that are limited in their growth and do not metastasize are called **benign**.

Cells become cancerous (also called **malignant**) only after sustaining damage to multiple genes. Damage to multiple genes often arises as a result of large-scale rearrangements or deletions of large stretches of DNA. There is no one set of genes that are consistently damaged in cancer. Different cancers can have quite different sets of damaged genes. Two different sorts of genes must be affected: **Oncogenes**, which promote cell division and migration, must be turned on, and **tumor suppressor genes**, which provide defenses against uncontrolled growth, must be turned off. The genes most commonly damaged in cancers include those involved in signal transduction, growth control, apoptosis, cell movement, and DNA damage recognition and repair. Each of these changes contributes to the damaging phenotype of the cancer.

Damage to many oncogenes results in derangements of signal transduction pathways, which cause the cell to respond as if a signal were either permanently on or off, regardless of the molecular environment. When a cell acts as if its growth or migration signals are always on, particularly when it also ignores any apoptotic signals, cancer can result. Cancer cells can also produce inappropriate signals, including ones that cause the body to grow new blood vessels to feed a growing tumor, or even growth factors that further exacerbate the tumor's own proliferation.

Most cancers are monoclonal, meaning that they initially arose from a mutation in a single cell. As a cancerous cell replicates, dysfunctional DNA damage detection and repair machinery can cause its progeny to accumulate additional mutations, and become increasingly deranged. As cancers grow, they become less differentiated, losing the structures and functions that characterize normal cells. The increased rate of mutation in cancer cells causes the production of many variants. Although most of these are neutral or deleterious, some additional mutations improve the ability of the cancer cells to grow and spread within the body.

The molecular basis of the processes of invasion and metastasis, the hallmarks that distinguish cancer from benign tumors, are increasingly well understood. Cancer cells tend to stop expressing cell adhesion molecules, such as the cadherins, allowing them to move relative to the other cells in their environment. The invasion of nearby tissues often involves dissolving barriers through the secretion of proteolytic (protein cutting) enzymes into the extracellular matrix, either by the cancer cell itself or by its signaling to nearby fibroblasts. The expression of proteins called integrins may also play a role in allowing cancer cells to invade other areas. Once cancer cells reach the blood or lymphatic system, they disseminate throughout the body. They may lodge in distant locations by chance, or through specific interactions with other cell types, also mediated by integrins.

Although immune surveillance can target abnormal self-cells for destruction, cancer cells have multiple ways of avoiding this fate. Sometimes they stop expressing the MHC antigen presentation molecules that immune cells use to assess self-cells. They can also produce immunosuppressive cytokines that cause generalized downregulation of the immune system.

Tumor suppressor genes are built-in controls to prevent uncontrolled proliferation. Some of the most important of these are the DNA damage recognition and repair proteins. When DNA damage that cannot be immediately repaired is detected, a protein called p53 halts the cell cycle. Either a longer-acting mechanism can repair the DNA and the cell cycle restarts, or the cell is permanently prevented from dividing, called entering **senescence**. Damage to p53, which is sometimes called the guardian of the genome, is one of the few genetic changes that appears in the majority of cancers.

Senescence is closely related to aging. Organs deteriorate as more and more of their cells enter senescence. Experimental studies that increased the amount of p53 made by cells in mice showed that these animals had an increased resistance to cancer, but unfortunately at the cost of premature aging. Though this balance may be manipulated to improve the treatment of cancer,[7] it suggests that aging might be one of the body's methods of protecting against cancer.

There are other interesting relationships between cancer and aging. Telomeres, the long sequences of repeating heterochromatic DNA at ends of chromosomes, are another means by which bodies limit the division of somatic cells. During division, DNA polymerase does not copy all the way to the end the chromosome, so each time a cell divides, its telomeres get a little shorter. When telomeres become too short, their structure changes (called uncapping), and the cells enter senescence. Many aging-related diseases are linked to shortened telomeres, as is aging itself. Of course, some cells, such as stem cells, must divide extensively during the lifetime of the organism. These few cell types express an enzyme called telomerase, which extends the telomeres after every cell division. Germ cells also express telomerase. Most cancer cells also express high levels of telomerase, demonstrating that they have escaped another of the body's layers of defense.

The hard-won and still incomplete molecular understanding of cancer has led to some new treatments. Traditionally, cancer has been treated either by surgery (which is difficult, since even a single cancer cell left behind can cause a recurrence of the disease), or killing all fast-dividing cells by drugs or radiation (which has severe side effects, since many normal cells in the body are also affected).

New anticancer drugs specifically inhibit the functioning of oncogene proteins, particularly those in signal transduction pathways, adding to the effectiveness of existing therapies, but not yet replacing them. For example, trastuzumab (Herceptin) is an antibody that binds to and blocks a growth factor receptor tyrosine kinase called her2/neu. More than 20% of breast cancers express a great deal of this receptor, and, in combination with standard chemotherapy, the drug gives patients with this otherwise very aggressive tumor a nearly 80% 1 year survival rate. However, there are serious cardiac side effects in about 5% of the population. Perhaps an even more important new cancer drug is imatinib (Gleevec), which is a small-molecule inhibitor of several signaling pathway kinases. It has become the first-line therapy for the cancer of the white blood cells called chronic myelogenous leukemia, and also has shown promise in the treatment of a variety of other cancers. Many new drugs are under development, and the treatment of cancer is likely to change substantially in the next generation.

7. See, for example, Mendrysa and Perry, Tumor suppression by p53 without accelerated aging: Just enough of a good thing? *Cell Cycle*, 5; 7 (2006, Apr.), 714–717.

9.7 Drug Discovery

Pharmacotherapy is a large and growing part of contemporary medical practice. Ingesting specific substances to cure a disease or heal an injury is a practice as ancient as humanity, and there is evidence that many species of animals also seek out and consume substances for their healing properties. The last hundred years or so have seen a tremendous explosion in the number and effectiveness of pharmaceuticals. While historically drugs were discovered by testing many naturally occurring substances, increasing knowledge of biology has opened the door to the invention of new drugs without natural analogs, based on understanding of the structure and function of biological molecules.

The basic process of drug discovery involves a series of steps. The first step is to identify a target on which the drug is intended to act. Such targets were traditionally defined in physiological terms, such as inhibiting the growth of a colony of microorganisms, or decreasing blood pressure in an animal. With increased molecular understanding, targets are now generally defined as particular molecules whose activities are to be modulated, for example, finding an **agonist** of a particular receptor. An agonist is a compound that activates or strengthens the function of a biological molecule; an **antagonist** is one that inhibits or weakens function. Drug targets are defined based on the understanding of the molecular basis of disease.

The next step in the drug discovery process is to identify a druglike compound that interacts with the target. What makes a compound druglike is a combination of factors that involve the practicality of administering it to patients (e.g., it should ideally be active when taken by mouth), possible toxicities, and the difficulty of manufacturing commercial quantities of it. Recently, Christopher Lipinski described the "rule of five"[8] that characterizes druglike compounds as those that have (1) not more than five hydrogen bond donors, (2) not more than ten hydrogen bond acceptors, (3) a molecular mass of under 500 Daltons, and (4) a measurement of hydrophobicity, called cLogP, of less than 5, meaning that the compound is fairly soluble in water. These rules have since been refined, but the basic idea is well accepted. Drugs of this type (as opposed to, say, antibodies or proteins administered as drugs) are generally referred to as small molecules.

8. Lipinski et al., Experimental and computational approaches to estimate solubility and permeability in drug discovery and development settings, *Advances in Drug Development Reviews*, 46 (2001), 3–26.

The most common way to find a small molecule that affects a target is through screening, that is, to test as many candidates as possible. One of the key issues in drug discovery is finding druglike compounds that are worth screening. A traditional and still powerful approach is to screen natural products. For example, the substance in the potent anticancer drug paclitaxel was discovered by a National Cancer Institute natural product screen of an extract from the bark of the yew tree. The secondary metabolites of many plant species have been found to have medically useful properties, and an enormous number of species, some with a history of use in traditional medicine, have never been screened.

Medicinal chemists have also developed methods for creating collections of synthetic druglike small molecules. Novel, druglike molecules are produced through **combinatorial chemistry**. In combinatorial chemistry, large sets of functional groups are combined in systematic ways to produce enormous numbers of new compounds. Millions of these synthetic compounds are now in the libraries of pharmaceutical companies, and a significant proportion of research resources in the pharmaceutical industry is spent screening them against targets. A contemporary industrial **high-throughput screening** system can test roughly 100,000 compounds a day. The anticancer drug imatinib, discussed earlier, was discovered this way.

Compounds that interact with a target are called hits. Once one or more hits are identified, they are more carefully assessed for their potency and specificity. Potent compounds bind to their targets even at low concentrations; a good drug candidate will bind to its target even in concentrations of a micromolar (one millionth of a mole) or less. Specific compounds bind only the target, and no other biological molecules. Nonspecific compounds are undesirable since they are likely to cause side effects. A molecule that binds potently and specifically to a target is called a **lead compound**.

Screening is not the only mechanism that can be used to identify lead compounds. If the structure of the target molecule is known, it is sometimes possible for medicinal chemists to use that knowledge to design a compound to interact with it. The region of the target that is of interest, for example, its active site or a region of allosteric regulation (see section 5.1.1), is used to form a **pharmacophore**. A pharmacophore defines the characteristics that a drug must have to interact with the target, generally its shape and the distribution of charges on its surface. Chemical and computational approaches have been defined to find molecules that fit a pharmacophore, creating new lead compounds. The ACE inhibitors described earlier were invented this way.

Once a lead compound has been identified, medicinal chemists go to work on optimizing it. Optimization involves creating variants of the compound intended to increase its potency (most drugs have picomolar—one billionth of a mole—binding affinities) and to test for and address possible nonspecific interactions, particularly with molecules known to be responsible for toxicities. For example, many drugs with cardiac toxicities bind to a potassium channel protein called hERG that plays a key role in the action potential of the cardiomyocyte. Even modest affinity for hERG can cause serious problems, so to progress in the development of a drug, variants must be found that exhibit no hERG binding at all.

Optimized compounds are then tested in animal models of the disease to ensure that the molecular activities translate into safety and effectiveness in intact animals. This is called **preclinical** research. Issues that arise here can often be addressed by further optimization. The progression from in vitro chemistry to in vivo models (which can involve a variety of species), and from optimization to animal testing is not always linear, and the processes of testing and optimization are often iterated.

When a specific compound that shows good safety and effectiveness in animal models has been developed, an application is filed to regulatory authorities (in the United States, this is the FDA) for an investigational new drug (IND). IND applications contain information about the preclinical results, the chemistry and manufacturing of the compound, and the proposed protocols for testing on human beings. When the compound is approved as an IND, **clinical trials** can begin. Clinical trials have three phases, and passage from one to the next requires additional regulatory approval. In phase I, small quantities of the drug are given to a small number of people (typically between 20 and 100) to assess initial safety and to determine the appropriate dosing levels in humans. Often, phase I trials are conducted on healthy volunteers or, sometimes in the case of cancer and AIDS, on patients who have failed all other therapy. Phase II trials moderately increase the number of participants (up to about 300), and begin to test for efficacy in ill patients. Phase III trials are the final tests necessary before general approval of the drug. These studies often involve thousands of patients, and are aimed a definitive assessment of the safety and efficacy in comparison to the best existing treatment. Most of these studies involve a double-blind, randomly controlled trial, meaning that neither the patients nor the doctors treating them know which treatment they are receiving, and the choice of who gets which treatment is made randomly. Clinical trials are slow (often it takes a long time to recruit enough of the right sort of patients, and it is important to understand the action of the drug over time) and expensive. The process of going from a lead compound to an approved drug can take 10 or more years, and cost hundreds of millions of dollars.

Many drugs that appear to be promising in preclinical studies fail to win final approval, usually due to concerns about safety. Drug toxicities can arise through a number of mechanisms. Exogenous compounds are metabolized by a set of detoxification pathways, mostly in liver cells. These pathways transform foreign compounds so that they can be excreted from the body. There are a large number of these enzymes; one important family, the cytochrome P450s (CYPs), includes more than 60 proteins that act on an enormous variety of substrates. The chemistry of these detoxification pathways is complex, since they must be able to handle any compound that can be ingested. As a result, these enzymes are less specific, and the set of intermediate compounds produced during detoxification can be quite diverse. Sometimes the metabolism of a drug produces a compound that is toxic, even though the drug is not. (It is also sometimes the case that the effective compound isn't a drug itself, but a metabolite produced by these pathways.) Another reason that a drug may turn out to be unsafe is that a portion of the population has a genetic variation in receptors, detoxification pathways, or elsewhere that causes the drug to have a different effect in those people. The study of **pharmacogenomics** attempts to understand how genetic variation interacts with drug action.

9.8 Molecular Medicine

Molecular biology still plays a modest role in medicine, primarily through its contributions to drug discovery. Medical practice does use a few assays that involve molecular biotechnology, such as the use of antibodies, but it is not a large part of most medical practices. Yet the promise of molecular medicine is great, and a great deal of effort is being expended to make it more of a reality.

There are many reasons that molecular biological insight has had a limited impact in medicine thus far. Our understanding of molecular biology is still new and relatively limited. Given the long development time for new drugs, currently available medicines will always reflect the state of knowledge from at least a decade ago. Recent applications for approval of new medicines show a significant shift from chemical approaches (traditional small molecule drugs) to more biological ones (such as antibodies and ribozymes).

Although our knowledge of molecular biology in general has grown rapidly, much of the specific knowledge needed by medicine is still elusive. Consider that any two human genomes are about 99.9% identical, so the differences among us, particularly the differences with medical significance, are relatively small. Genomes are large, so despite the genomic similarity among all people, there are hundreds of millions of SNPS and an unknown number of other polymorphisms, any combination of which might play some role in human disease. Figuring out which of these are medically important is an enormous task, and the focus of much current research.

Sometimes the knowledge of what gene or genes contribute to a disease is not enough to make a medical difference quickly. For example, muscular dystrophy was one of the first diseases that was traced to a defect in a specific gene, by Louis Kunkel in 1986. The identification of the gene led almost immediately to a genetic test that prospective parents could use, but more than 20 years later, the molecular understanding has not led to effective treatment for people with the disease. One of the reasons for the difficulty is that mice that are given defective versions of the gene do not develop the disease, making it hard to study.

Despite these challenges, the pharmaceutical industry has adopted molecular methods wholeheartedly, and much progress has been made in a relatively short time. In time, increasing appreciation of the molecular structures and functions of life will almost certainly bring dramatic improvements in medical care, and in other sorts of biotechnology.

9.9 Suggested Readings

There are countless books about medicine for the nonspecialist, but perhaps the best way to get a sense of what medicine is really like is to enroll in a mini-med school. These programs typically involve a free public lecture series that tracks the content of the regular medical curriculum, but is aimed at people with minimal scientific backgrounds. The program was started at the University of Colorado in 1989 and has spread to more than 70 medical schools around the world. A good source of information is the U.S. National Institutes of Health: http://science-education.nih.gov/minimed. It is also possible to read the same textbooks that are used to train clinicians, but without trying to commit the contents to memory; popular books include *Davidson's Principles and Practice of Medicine* (20th ed., Churchill Livingston, 2007), Kumar and Clark's *Clinical Medicine* (6th ed., Saunders Ltd, 2005), or for those who prefer a more visual presentation, Forbes and Jackson's *Color Atlas and Text of Clinical Medicine* (3rd ed., Mosby, 2005).

Nontechnical books on drug discovery are harder to find. Bartfai and Lees's *Drug Discovery: From Bedside to Wall Street* (Academic Press, 2005) is an accessible account of the process based on detailed examples of successes and failures. Andy Kessler's *The End of Medicine: How Silicon Valley (and Naked Mice) Will Reboot Your Doctor* (Collins, 2006) is an entertaining look at how technology is likely to change the practice of medicine, although the book is a little too glib in some regards.

10 Molecular Biotechnology

The previous chapters describe the current state of knowledge in molecular biology. It is important to keep in mind that all science is provisional, and some of the statements made here may turn out to be wrong. Biology is in a particularly creative period now, with long-held ideas (e.g., eyes evolved independently multiple times) being overthrown by new data (conservation of the pax-6 eye morphogen across the metazoa) with startling frequency. In order to properly assess the evidence that supports current knowledge, it is important to understand its experimental underpinnings and the technology that makes the study of molecular biology possible. Although this brief survey isn't enough to develop a sense of which findings are most reliable and which are more dubious, knowing about these technologies is the first step toward a more nuanced appreciation for their strengths and weaknesses.

This same technology also opens the door to remarkable possibilities in bioengineering, designing life forms and biological interventions, perhaps even in the human germline. The practice of medicine and pharmacology is also changing, to adopt some of these molecular techniques in diagnosis and develop new therapies.

The experimental techniques used in molecular biology seem almost as diverse and elaborate as the living things being studied. Some of the most widely used instruments and methods are described in this chapter, but this is just a small sampling of the tools and techniques used in molecular biology laboratories.

10.1 Molecular Instrumentation

Much of the biology presented in the previous chapters describes a particular molecule catalyzing a particular reaction, or interacting with another particular molecule, or being present at a particular time and place. Since most molecules cannot be observed directly (even with the most powerful optics), the evidence for these facts has to be gathered through molecular instrumentation. Some understanding of how the instrumentation works is important in assessing the credibility of statements in the literature and understanding the limits of what can be known.

10.1.1 Measurements of Mass and Charge

The most basic measurement that can be made regarding a molecule is its mass. Measurements of the molecular mass of a pure compound can be made with very high accuracy, and it is possible to distinguish between compounds with very slightly different masses, even when they are part of complex mixtures. Molecular mass is measured in daltons, which is same as the unit of atomic mass used in the periodic table shown in figure 3.1.[1] If only one thing is known about a macromolecule, it will be the mass; it is common to see references to, say, a 60-kD protein, which means its mass is about 60,000 daltons. The most studied human protein, called p53, got its name for being a protein with a mass of about 53 kD.

The **centrifuge** is an instrument that accelerates a sample by spinning it at very high speed, causing its components to separate based on their density. This instrument is often used to isolate particular cells, organelles, or even macromolecules. A centrifuge that can generate the very great acceleration needed to separate macromolecules is called an ultracentrifuge. The rate at which the materials accumulate at the boundary, called the sedimentation rate, can be used to estimate the mass of the material, although not with great precision.

Gel electrophoresis is a more sensitive method for separating molecules by mass, and a more accurate method for estimating mass. A gel is a viscous polymer that forms a matrix through which macromolecules can move, but only slowly. An electric current is run through the gel, and the macromolecules are pulled through it based on their charge. To measure the mass using this technique, proteins are denatured by applying a detergent called sodium dodecyl sulfate, or SDS. Denaturing, in this case, means transforming a protein into an elongated conformation with a high negative charge. In this condition, the time it takes a protein to migrate through the gel is proportional to its mass. This widely used technology is called SDS-PAGE (PAGE stands for polyacrylamide gel electrophoresis). As shown in figure 10.1, proteins of a

---

1. Formally, it is one-twelfth the weight of the most abundant carbon atom (which contains 6 protons, 6 neutrons, and 6 electrons). A proton, a neutron, and a hydrogen atom all have masses of about 1 dalton.

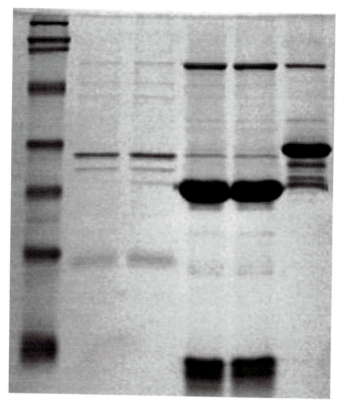

Figure 10.1
An SDS-PAGE gel. The column on the left shows a standard with compounds of known molecular weights. Five additional columns show the presence and absence of proteins with particular molecular weights in various conditions. Dark bands indicate the presence of high concentrations, light bands lower concentrations of protein. The absolute molecular weights can be inferred by comparison with the standard column on the left. This image is by Magnus Manske, taken from Wikimedia Commons, and is used under the GNU Free Documentation License.

particular mass will appear as a band in a column of a gel, and the abundance of a protein can be roughly estimated from the intensity of the band. One gel often shows several columns, each loaded with a sample from a different experimental condition, demonstrating the presence and absence of various compounds in each condition. Sets of closely spaced bands, such as can be seen in the far right column in figure 10.1, often indicate the presence of a protein with several different posttranslational modifications.

SDS-PAGE can be used to estimate the molecular mass of a protein by the use of a column loaded with a set of markers of known mass. This method, although approximate, was widely used until the era of inexpensive DNA sequencing. The mass of the p53 protein was estimated this way. However, the unusually large proportion of proline residues in p53 caused it to migrate through the gel more slowly than its mass would indicate; its actual mass turns out to be less than 44 kD, although the name stuck.

In complex mixtures, such as those obtained from a whole cell, there are likely to be many proteins whose molecular masses are very similar, despite having different sequences. To further separate these proteins, a pH gradient can be applied at 90 degrees to the electric charge, separating the proteins both on mass and on changes in charge related to pH (called the isoelectric point). This technology is called two-dimensional electrophoresis, or 2D-PAGE. Figure 10.2 shows a highly analyzed 2D PAGE image from a human kidney sample. Each spot is a protein, and many of the spots have been identified and labeled. PAGE, particularly 2D-PAGE, can be used to purify proteins (each spot is likely a pure compound), which can then be blotted or cut out of the gel for further experimental work, such as identifying the protein. A **Southern blot**, named after Edward Southern, is a means of transferring a purified DNA fragment from a gel. Northern, Western, and similarly named blotting techniques are variants for identifying different sorts of compounds (RNA, proteins, etc.) from gel bands or spots.

The most sensitive instrument for measuring molecular mass is called a **mass spectrometer**, often "mass spec" for short. These instruments can simultaneously provide information about the mass of all the different compounds in a mixture. Mass spec requires that the molecules it analyzes have a net charge, so various techniques are used to ensure that the molecules that are fed into the mass spec are ionized. The spectrometer then accelerates the ions using an electromagnetic field, and sends them at high speed through a vacuum chamber toward a detector. If the force exerted on the molecules is fixed, then the acceleration each experiences will depend on its mass, based on Newton's second law. Different accelerations mean compounds will arrive at the detector at different times. The force exerted on a compound actually depends on its charge, so the output of a mass spec is really a measurement of the mass-to-charge (m/Z) ratio. In general, it is possible to either know the charge of all the compounds or to ensure that the charge is fixed (usually at 1). The output of a mass spectrometer, shown in figure 10.3, lists the masses of the different compounds in the input mixture. Each peak represents a separate mass. High-resolution spectrometers can resolve differences in mass of a fraction of a Dalton, showing even isotopic differences (e.g., a difference of a single neutron in a 100-kD protein). The heights of the peaks indicate the amount of each compound in the sample. Those peak heights are a good indication of relative abundance of compounds, although it remains difficult to map them to an absolute concentration.

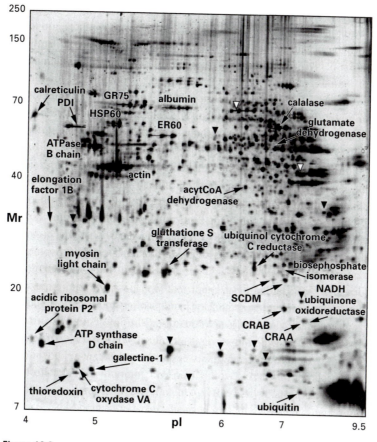

Figure 10.2

A 2D-PAGE gel image, with many of the proteins labeled. The vertical access is the mass range, and the horizontal axis the isoelectric point. Groups of nearby points with lines drawn through them (e.g., for albumin) indicate a protein with its posttranslational modifications. This image is from the Swiss 2D gel database, which freely gives permission for the use of its images in publications.

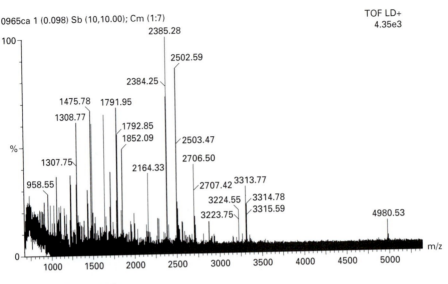

0965ca 1 (0.098) Sb (10,10.00); Cm (1:7)

TOF LD+
4.35e3

Figure 10.3
A mass spectrum. The peaks are labeled with the mass of the compounds in the mixture. The horizontal axis is mass to charge ratio, the vertical indicates relative abundance. Image courtesy of Katheryn Resing.

Identification of proteins from small samples and from complex mixtures is an important and difficult task. Although expression microarrays (described later) are able to provide a great deal of information about which proteins are being synthesized at any particular time, the long and active life of proteins after they leave the ribosome (e.g., through posttranslational modifications) means that direct protein assays are also important. The approaches used to identify, and sometimes quantify, all of proteins in a complex mixture are called **proteomic** technologies, and mass spec plays a central role. Proteomic mass spec is a complex process, but the essentials are straightforward. Although many proteins have the same mass, the proteins can be broken up (digested) by an enzyme that splits them at specific amino acids, often producing an identifying "fingerprint" of peptide masses that can be matched to a database computationally. Proteins that remain ambiguous in this approach can sometimes be resolved by dividing the initial protein mixture into fractions with different chemical characteristics. A still more elaborate approach, called **tandem mass spec**, hooks two spectrometers together. After the mass of a peptide is determined by the first mass spec, the peptide is further broken up and passed to a second mass spec. The fragment masses in the second mass spec can be used to determine part of the amino acid sequence of the peptide, which in combination with the data from the first instrument can quite specifically identify the original proteins, as well as many posttranslational modifications.

Mass spectroscopy is also used to separate and identify lipids and small molecules in biological samples. In analogy to proteomics, the ability to identify or characterize all of the lipids in a sample is called **lipidomics**. For all of the small molecule metabolites, the equivalent process is called **metabolomics**. Since many different small molecules have the same mass, metabolomics is sometimes carried out with magnetic resonance spectroscopy, described in the next section.

10.1.2 Macromolecular Structure Determination

The molecular mass of a protein is useful for identifying it, but does not provide much information about its structure. Recall that the three-dimensional structure of a protein determines its activity, and is of great importance in both understanding native function and designing pharmaceuticals. Determining the structure of a protein remains a difficult, time-consuming, and expensive task in most cases, although rapid technological progress in reducing these challenges has been made in recent years. Two methods are widely used: X-ray crystallography and nuclear magnetic resonance spectroscopy (NMR). X-ray crystallography starts by creating a crystal of a protein, which orders a substantial quantity of pure protein molecules in a regular pattern. High-energy X-rays are passed through the crystal and diffracted by the electrons, producing an image such as the one seen in figure 10.4. The intensity (amplitude) of each point is almost enough data to determine the structure of the protein; however, the relationship between the X-ray waves that created the points (phase) is also needed, which cannot be retrieved directly from the image. Various methods involving either manipulating the protein (e.g., by attaching metal ions to it) or using relationships in previously solved structures are used to determine the phase, and therefore the electron density throughout the protein, and hence its structure. In any case, using X-rays to determine the structure of a protein requires growing a crystal of that protein, which is often the most difficult aspect of crystallography and is not always possible.

NMR, the same technology that is behind the MRI images used in medical diagnosis, can be used to determine the structures of some proteins. Certain types of NMR experiments provide information about which atoms (of different elements) are near each other in the protein, and precisely what those distances are. Different experiments interrogate different types of atoms, and different sorts of relationships between them (e.g., through a bond, or just being nearby in space). These experiments can be done in rapid sequence and

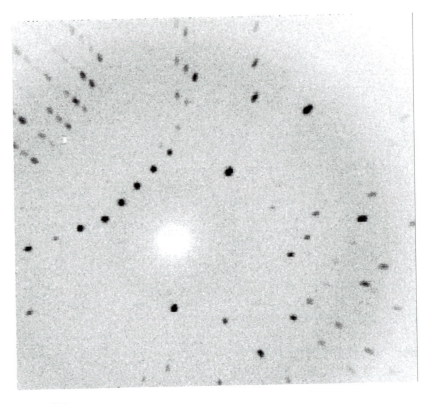

Figure 10.4
A portion of the image produced by X-ray diffraction through a crystal of myoglobin. Image from Brookhaven National Laboratory.

on a sample of protein dissolved in water. With enough information about these relationships, it is often possible to reconstruct the entire structure of the protein from these local distances, particularly for proteins that are not too large. One of the major advantages of NMR is that it doesn't require crystallizing the protein; a major disadvantage is the limitation on the size of the protein that can be solved. NMR can also be used to identify small molecules in solution, an application called **magnetic resonance spectroscopy** (**MRS**). This technique has been used in metabolomics.

Both NMR and X-ray crystallography can be applied to nucleic acids as well as proteins (in fact, the double-helical structure of DNA was determined by X-ray crystallography in the 1950s). An increasing number of structures of interacting sets of molecules have been determined, such as the structure of a transcription factor bound to DNA or even more complex molecular machines. Also, the structure of a protein is not always fixed; consider a receptor that changes its conformation when it binds a ligand, such as the receptor tyrosine kinases described in section 7.2. NMR experiments can capture motion, and have been producing a more sophisticated view of the changes in protein structure that occur as part of normal functioning. A database of all the publicly known macromolecular structures, called the protein databank (PDB) is at http://www.pdb.org.

The determination of the amino acid sequence of a protein (its primary structure) was once an important experimental technique, but the advent of rapid and inexpensive nucleic acid sequencing technology (described in section 10.1.5) has replaced it completely.

10.1.3 Assays of Molecular Activity

Determining the function of a macromolecule is often more challenging than determining its structure. The most basic approach is to directly assay the proposed molecular function. For example, demonstrating that an enzyme catalyzes a particular reaction, in isolation from the rest of the organism. An experiment that is carried out in isolation from an organism is called **in vitro**. Such experiments clearly demonstrate that the compound studied has the function claimed, but do not prove that it is the only function, nor what function the compound carries out in a living system. Only experiments done in an organism, called **in vivo**, can demonstrate that. A typical experiment to assay the function of a protein in vivo is to demonstrate that the function does not occur when the protein is removed, and is restored when purified protein is added back.

It is possible to remove a protein from an organism completely and permanently by removing or damaging the gene that codes for it, called doing a **knockout**. Knockout experiments can provide clear evidence of the biological processes in which a gene product plays a role. The method for creating a knockout is described in section 10.2. Another, more targeted, way of removing a protein is to use RNA interference (described in section 5.2.2). However, RNAi sometimes lets a small amount of protein be produced, so these experiments are sometimes called **knockdowns**, rather than knockouts.

The sort of organism that arises when a particular gene is knocked out or otherwise modified is a **mutant phenotype**, which can be contrasted to the normal, or **wild type**, phenotype, providing information about the function of the knocked out gene. Many genes, particularly in fruit flies, are named after the phenotype of knockout mutants. This can be confusing, since the wild type activity will be related to the opposite of the mutant phenotype. For example, the protein called "dunce" plays a crucial role in learning and memory, and the protein "wingless" is a morphogen that influences wing development. A significant number of genes are required for development, and so knocking them out results in the early death of the embryo; such genes are called **essential**. To assay the effect of the loss of these genes in adult cells, it is possible to design **inducible knockouts**, where some other signal (like a drug) can be used to shut off production of a gene later in life. Another experiment, called **synthetic lethality**, is used to determine genes that interact with each other in essential cellular processes. These experiments, generally done in yeast, identify pairs of genes that are not individually essential, but are lethal when both are knocked out simultaneously.

An important kind of molecular function is to bind to other molecules. Assays that directly measure the binding affinity of one protein to another are a staple of molecular biology. It is also possible to rapidly screen a large number of proteins to see which binds to a particular target. The target, a small molecule or another protein, is immobilized and the to-be-screened molecules are washed over it. The amount of any of the screened molecules that remains after rinsing can be translated into a measure of the **binding affinity** of the two compounds. An in vivo version of this experiment, called **yeast two-hybrid**, is capable of efficiently screening an enormous number of proteins for binding affinity to a protein target. This experiment takes place in a yeast cell, but can test for binding affinity of proteins from any organism.

10.1.4 Distribution of Molecules Through Space and Time

There is a difference between the biochemical function of a protein (e.g., as a catalyst) and its functioning in the life of an organism. What work the protein does in an organism depends on where and when the protein appears, and when and how it interacts with its partners. Techniques that provide information about the distribution of a protein and other molecules it interacts with are therefore important experimental tools. Often the question of distribution is assayed by making concentrations of the protein visible under a light microscope.

The traditional approach to making particular biological entities visible is to stain them with a dye that is specific to them. Many stains have been developed that target biologically important features in cells and tissues. However, traditional stains are generally not so specific they can pick out an individual protein or stretch of a nucleic acid, which makes them less well-suited for molecular biology. More recently, biologists have borrowed from the adaptive immune system a biologically based approach to tagging specific proteins.

Antibodies recognize and bind to particular peptides with high specificity, and form the basis for many important experimental methods. Pure antibodies (called monoclonal) that recognize and bind to an enormous variety of peptides are available commercially. Antibodies to new proteins can generally be produced by injecting an animal with that protein and harvesting the resulting plasma cells, which can then be maintained in a type of cell culture called a hybridoma. Antibodies can be used to capture a pure protein from a mixture, called **immunoprecipitation**, or can themselves be tagged with a dye for staining and visualization of the protein they recognize.

Antibodies can also have problems, including imperfect specificity. More effective yet is to attach a tag that can be rendered visible directly to the protein of interest. One way to do that is to change the gene that codes for the protein (see section 10.3 for how this works) to add a part that can be made visible. Often this involves fusing the relatively small **green fluorescent protein** (**GFP**) from the jellyfish *Aequorea victoria* to the protein of interest. When illuminated with blue light, GFP glows green, giving a very clear signal of the location of the protein of interest in a microscopic image. Figure 10.5 shows a neuron with GFP fused to a protein called AKAP79; the protein can clearly be seen in the cytoplasm, and to localize to dendritic spines (the receiving side of synapses). Although GFP is perhaps the most widely used, other proteins also fluoresce different colors (yellow, blue, etc.) and can be combined to produce images that show the location of multiple proteins simultaneously. Other sorts of stains and dyes that color cell components possessing particular chemical properties (such as the relatively acidic nature of the nucleus) can also be combined with these, producing beautiful and informative images (for example, figure 6.1). The use of fluorescent tags on proteins can be refined in many different ways. One powerful technology, called fluorescent resonance energy transfer, or FRET, fluoresces only when two particular tags are brought into very close proximity, making it an ideal assay for determining where and when two particular macromolecules interact with each other.

10.1.5 Nucleic Acid Instrumentation

The instruments and techniques that generate information about protein structure and function have done a great deal to advance knowledge of molecular biology in the last hundred years. However, the most revolutionary advances in knowledge have arisen from instruments that manipulate nucleic acids.

At the basis of all of this technology is an instrument called a **DNA sequencer**, which can determine the sequence of nucleotides in a nucleic acid molecule. The first methods, developed in the 1970s by Frederick Sanger,

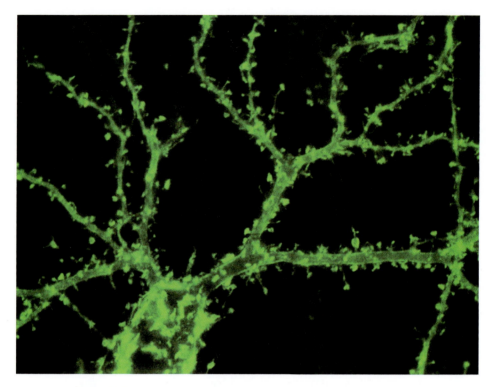

Figure 10.5
The fusion protein AKAP79-GFP expressed in a rat hippocampal neurons. Image courtesy of
Mark Dell'Acqua, PhD, University of Colorado at Denver.

Wally Gilbert, and others, worked by producing fragments of the targeted
DNA molecule with a known nucleotide at one end. Those fragments could
be separated on an electrophoresis gel, generating a ladder of bands, each one
nucleotide longer than the previous, and each with a known nucleotide at the
end, as shown in figure 10.6. The requirement that fragments that differ by
only a single nucleotide be clearly distinguishable puts a limit on the length
of the sequence that can be determined at once with this method. Modern
versions of this technology use four-color fluorescent labels, capillary electrophoresis (no gels), and computer-controlled equipment to achieve a similar
result, much more quickly and inexpensively. While the length limits have
been extended somewhat, sequencing by this method is still confined to continuous fragments of about 1,000 nucleotides.

Even relatively small genomes have several million nucleotides; humans
and many other organisms have billions of nucleotides in their genomes. The
shotgun sequencing method was devised in order to reliably reconstruct the

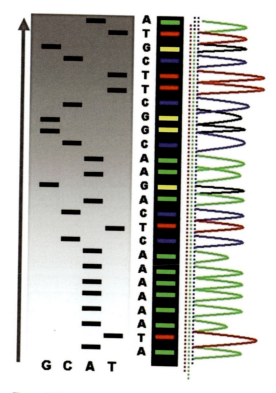

Figure 10.6

A schematic of a Sanger sequencing gel. On the left, a drawing of an idealized sequence ladder. Each column in the gel contains sequences that end with a particular nucleotide (created by adding chain-terminating nucleotides that lack the 3' OH group). The sequence of bands shows each length of a DNA fragment that ends in that nucleotide, allowing the overall sequence to be read off up the gel. Labeling the ends of the fragments with fluorescent dye makes it possible to determine the sequence without gels, the basis of modern DNA sequencing machines. This image was created by Abizar Lakdawalla and is used under the Creative Commons Attribution-ShareAlike license.

sequence of a large stretch of DNA from many short sequences. With enough overlapping short sequences, in most cases, it is possible to figure out the original order of the fragments computationally. This computation is most difficult when the DNA consists of very long regions of short repeating sequences, where the method has trouble determining the exact number of such repeats. While euchromatin, which contains most of the protein-coding regions, is generally straightforward to sequence, the heterochromatin of the centromeres and the telomeres is much more difficult.

New DNA sequencing technologies and improvements in the efficiency of existing instruments have rapidly driven down the cost of sequencing, which have fallen more than 100-fold in the last fifteen years. It appears likely that the price of sequencing will continue to fall; the goal of one U.S. government program is to drive the cost of sequencing below $1,000 for a sequence the size of the human genome (3 billion base pairs). As of this writing, nearly 100 billion base pairs of DNA sequences have been deposited in the GenBank public database,[2] and the complete genomic sequence is available for 27 eukaryotic genomes, with 329 more in progress. Complete genomes are currently available for 564 bacteria and nearly 3,000 viruses.

DNA sequencing technology also makes possible the determination of protein-coding regions and many other features of interest. Messenger RNA can be purified from cells, for example, by purifying RNAs that have a poly-A tail. These mRNAs can be used as a template to produce **complementary DNA** (or cDNA), which can then be sequenced and conceptually translated to determine the protein sequence. Comparing a cDNA sequence to the genomic DNA sequence from the same organism shows where the exon and intron boundaries are. Multiple cDNAs arising from the same genomic region provides evidence of alternative splicing. Sequencing the same region in many individuals identifies SNPs and other important sequence polymorphisms.

Often DNA molecules of interest, such as mRNA produced in a single cell, are present naturally only in very low concentrations. However, if even a part of sequence of the rare DNA is known, a technique known as the **polymerase chain reaction**, or **PCR**, can amplify that sequence, creating a large number of copies very rapidly. Each round of PCR doubles the amount of DNA present, so it is possible to use the method to quantify even tiny concentrations fairly precisely.

It is also possible to synthesize arbitrary DNA sequences. The chemical reactions necessary to produce a DNA molecule with any given sequence are straightforward, and automated DNA synthesis instruments are available. Short DNA sequences can be created cheaply and easily. However, accurately creating long DNA sequences is done by splicing short sequences together, which is still relatively difficult and therefore expensive. Producing a thousand-base-pair sequence from scratch costs nearly $1,000, and can take many days, although this area is ripe for technical advances.

One very important property of nucleic acid sequences is their propensity to bond to exactly complementary sequences, called **hybridization**. Even a single base-pair mismatch will cause a significant drop in a sequence's affinity for another sequence. Hybridization, combined with knowledge of genomic

2. http://www.ncbi.nlm.nih.gov/Genbank.

Figure 10.7
Results from an experiment showing the expression of thousands of genes on a single GeneChip®
probe array. Image courtesy of Affymetrix.

sequence, is the basis for a powerful set of instruments, collectively known as
microarrays.

Microarrays, shown in figure 10.7, fix fragments of DNA with known
sequence (called probes) to positions in a large matrix on a glass slide or a
silicon-based chip. A sample containing many unknown sequences is tagged
with a fluorescent label and then washed over the array. The unknowns hybridize to complementary probes in the array. Positions on the array that fluoresce
indicate that a sequence among the unknowns matches the probe at that location. The brightness of the fluorescence is proportional to the concentration of
that sequence in the mixture. If the arrayed sequences correspond to genes and
the unknowns are all the mRNAs produced by a cell, the chip assays how

much of each gene is being transcribed in the cell. This technique is called an **expression array**. If the arrayed sequences correspond to SNPs or other allelic variants and the unknowns are DNA sequences for those genes from a particular individual, the chip assays which variants that individual carries. This technique is called a **genotyping array**. A particularly elaborate approach, called **chip-chip**, starts by chemically locking down all of the transcription factors bound to DNA in a cell, then breaks up the DNA into small pieces, extracts transcription factors of interest and the DNA sequences they are bound to through use of antibodies, and then determines the sequence of the bound DNA through a microarray. These experiments provide specific information on all of the positions in the genome that are bound by a particular transcription factor, and therefore about the genes it regulates.

The techniques made possible by inexpensive DNA sequencing and hybridization-based assays are developing rapidly, and are likely to continue to produce new instruments of surprising and useful capabilities for many years to come. These technologies create very powerful instruments for interrogating the structures and functions of biological molecules. However, the ability to use these instruments to improve human knowledge of the workings of life requires careful (and often clever) experimental designs. One of the key aspects of a successful biological experiment is the right choice of experimental organism.

10.2 Model Organisms: Germs, Worms, Weeds, Bugs, and Rodents

Although millions of species exist in today's world, only a very small number of them are used widely in molecular biology experiments. The investigation of the workings of even a single organism is so complex as to take many dedicated scientists many careers' worth of time. Trying to study all organisms in great depth is simply beyond the capacity of modern biology. Furthermore, the techniques of biological experimentation are often complex, time-consuming, and difficult. Some of the most valuable methods in biological research are invasive, or require organisms to be sacrificed, or require observations over many generations or on large populations. For those and other reasons, much of this work is impractical or unethical to carry out on humans.

Alternatively, biologists have selected a variety of **model organisms** for experimentation. These creatures have qualities that make possible controlled laboratory experiments at reasonable cost and difficulty with results that can often be extrapolated to people or other organisms of interest. Of course, research involving humans can be done ethically, and in some areas of biomedical research, such as final drug testing, it is obligatory. Other research methods involve kinds of human cells that can be grown successfully in the laboratory. Not many human cell types thrive outside of the body. Some human cancer cells do grow well in the laboratory, and these, called **cell lines**, are an important vehicle for research.

Sometimes the selection of a new model organism can lead to advances in a field. For example, the use of a particular kind of squid made possible the understanding of the functioning of neurons because it contained a motor neuron that is more than ten times the size of most neural cells, and hence easy to find and use in experiments. Experimentally useful correlates of nearly every aspect of human biology can be found in some organism or another, but the following organisms are the ones predominantly used in molecular biology.

These organisms became popular for a variety of reasons. Some are particularly amenable to laboratory manipulation, having short generation times, thriving in the rather impoverished environment of a biology lab, or having features (such as the size of their embryos) that make them easier to analyze than other creatures. These organisms also represent a modest variety of the different branches of the tree of life, including a bacterium, a protist, a land plant, a simple animal, an arthropod, a vertebrate, and a mammal. Perhaps most important, once an organism gains acceptance as a model, the enormous amount of information generated about it, ranging from genome sequence to behavior, makes it ever more useful as an experimental subject. Since these organisms are so widely used in molecular biology, it is important to know a bit about each one.

Certain facts about these organisms, such as the size of their genomes or the number of genes, are proxies for the complexity of the organism. For comparison, the human genome contains 23 diploid chromosomes and a mitochondrial chromosome, together containing a total of 3,070,145,600 base pairs, with an estimated 28,749 protein-coding genes,[3] at least 9,986 of which are likely to have alternative splices,[4] as well as a minimum of 1,490 (and possibly as many as 5,000) non-protein–coding RNA genes.[5] For humans (and also mouse; see following text) the numbers of alternative proteins from a

3. These figures are taken from the U.S. National Library of Medicine's NCBI RefSeq genome analysis in August 2007 (http://www.ncbi.nlm.nih.gov/sites/entrez?db=genomeprj&cmd=Retrieve& dopt=Overview&list_uids=168 for human). The figures in the following text are also from the NCBI unless otherwise noted. The European EnsEMBL database lists 23,244 human genes.

4. Taken from the EST-Derived Alternative Splicing Database, http://www.genebee.msu.ru/edas.

5. Sam Griffiths-Jones, "Annotating noncoding RNA genes," *Annual Review of Genomics and Human Genetics*, 8 (2007), 279–298.

transcript and the number of non-protein–coding genes are fairly coarse estimates, and likely underestimate the true extent of these phenomena.

10.2.1 *Escherichia coli*

The ubiquitous intestinal bacterium *Escherichia coli* is a workhorse in biological laboratories. Because it is a relatively simple organism with fast reproduction time and is safe and easy to work with, *E. coli* has been the focus of a great deal of research in genetics and molecular biology of the cell. Although it is a bacterium, many of the basic biochemical mechanisms used by *E. coli* are shared by humans. For example, the first understanding of how transcription factors can turn genes on and off came from the study of a virus that infects these bacteria.[6] There are many strains (variants) of *E. coli*. The strain called O157:H7 is a particularly virulent pathogen, responsible for several lethal cases of food poisoning. The strain called K12 is widely used in the laboratory. K12 was one of the first free-living organisms that had its genome sequenced. It has a single circular chromosome of 4,639,221 base pairs, containing 4,243 protein-coding genes and 168 RNA-coding genes. The *E. coli* model organism database is at http://www.genome.wisc.edu/. Additionally, a remarkable amount of information about this organism is available in both human- and machine-readable form at http://ecocyc.org.

  E. coli is a common host organism for genetic engineering, producing all kinds of valuable proteins in great quantity at low cost. *E. coli* bacteria can produce the proteins coded for by most human genes. Industrial facilities use them to churn out large quantities of human proteins used as drugs, in cosmetics and for other purposes.

10.2.2 *Saccharomyces cerevisiae*

Saccharomyces cerevisiae is better known as brewer's yeast, and it has been used in the production of beer for millennia. It is another safe, easy to grow, short generation time organism. Yeasts are eukaryotes, so many of their cellular components and biochemical processes are homologous with those in people. Because these yeasts are so easy to grow and manipulate, and because they are so biochemically similar to people, many insights about the molecular processes involved in metabolism, biosynthesis, cell cycle, DNA repair, and

---

6. *A genetic switch: Gene control and the phage lambda* (1987) by Mark Ptashne was the discoverer's explanation of his work for a popular audience. It remains an excellent read twenty years later.

other crucial areas of biology have come from the investigation of yeast. The genome of *S. cerevisiae* is organized into 12 chromosomes containing about 12,166,000 base pairs, with an estimated 5,879 protein-coding genes and 418 noncoding RNA genes The model organism database for *S. cerevisiae* is http://www.yeastgenome.org.

10.2.3 *Arabidopsis thaliana*

The most important application of increased biological understanding is generally thought to be in medicine, and increased understanding of human biology has indeed led to dramatic improvements in health care. However, in terms of effect on human life, agriculture is at least as significant. A great deal of research into genetics and biochemistry has been motivated by the desire to better understand various aspects of plant biology. *Arabidopsis thaliana*, a weed commonly known as mouse-ear cress and related to mustard and cabbage, is the most widely used plant model organism. *Arabidopsis* makes a good model because it undergoes the same processes of growth, development, flowering, and reproduction as most higher plants, but its genome has 30 times less DNA than corn, and very little of the repetitive DNA found in many plant genomes. It also produces lots of seeds, and takes only about six weeks to grow to maturity.

Arabidopsis was the first plant with a completely sequenced genome. It has five diploid chromosomes, a mitochondrial chromosome, and a chloroplast chromosome together containing 119,707,902 nucleotides, one of the smallest genomes in the plant kingdom. The estimate at the model organism database in August 2007 is that it contains 27,029 protein-coding genes, 3,889 transposable elements and 1,123 noncoding RNA genes. The model organism database for *A. thaliana* can be found at http://www.arabidopsis.org/.

The complete genomes of three other plant species—rice, black cottonwood trees, and the forage legume barrel medic—are publicly available. Genome sequencing of several other agriculturally important species is in progress, including for corn, sorghum, tomato, potato, cassava, and lotus.

10.2.4 *Caenorhabditis elegans*

Caenorhabditis elegans is a free-living soil nematode; thousands can be found in a spadeful of dirt. It has proven to be an extraordinarily productive model for studying multicellular organisms. It is transparent, and the adult always consists of exactly 959 somatic cells. Not only is the complete anatomy of the organism known, but a complete cell fate map has been generated, tracing the developmental lineage of each cell throughout the lifespan of the organism. This map allows researchers to relate behaviors to particular cells, to trace the effects of genetic mutations very specifically, and gain important insights into the mechanisms of development and aging. Nearly one third of the cells of the adult worm are neurons, and it has proven a valuable model for relating neuronal connectivity to the various behaviors of the organism. *C. elegans* was the first multicellular eukaryotic genome to be completely sequenced. According to the model organism database at http://wormbase.org, it has six diploid chromosomes containing a total of 100,267,452 bases, with 20,146 protein-coding genes (2,376 of which have confirmed alternate splices) and 1,029 noncoding RNA genes (including 608 tRNAs, 8 rRNAs, and 137 miRNAs).

10.2.5 Drosophila melanogaster

Drosophila melanogaster, the black-bellied fruit fly, has been a staple of genetics research for more than a hundred years. The flies have short generation times, many easily visible genetically determined morphological characteristics (like eye color), and lay thousands of eggs at a time. Many of the critical discoveries in development, such as the identification of the homeobox genes, were made using *Drosophila*. Not only is the genome sequence of *D. melanogaster* known, but twelve other fruit fly species have also been completely sequenced, making possible important molecular comparative studies. *D. melanogaster* has six diploid chromosomes and a mitochondrial chromosome, containing a total of 120,401,377 base pairs, with an estimated 14,141 protein-coding genes. There are at least 4,817 fruit fly genes with alternative splices, and at least 1,040 non-protein–coding RNA genes. The model organism database can be found at http://flybase.net.

10.2.6 Danio rerio

Danio rerio, also known to tropical fish fans everywhere as the zebrafish, is a useful vertebrate model organism. It has a clear body and rapid generation time, in addition to a relatively small genome for a vertebrate. One of the reasons for the small size of the genome is that zebrafish genes have very few introns, another advantage for molecular biologists. *D. rerio* has 25 diploid chromosomes and a mitochondrial chromosome. Sequencing of the zebrafish genome is not complete as of this writing, but its genome consists of approximately 1,700,000,000 nucleotides, and it may have as many as 37,900 genes. The model organism database is at http://zfin.org.

10.2.7 *Mus musculus*

Despite the stereotype of a lab rat, the house mouse *Mus musculus* is a more widely used model mammal in the molecular biology community. As mammals, mice have organ systems and developmental pathways that are homologous to almost all of the human ones. Mice are susceptible to most of the same diseases and infirmities that plague people. Yet mice are relatively amenable to experimental manipulation. They reproduce rapidly, mature quickly, and are relatively easy to maintain. Mice have been experimental animals for a long time, and a wide variety of inbred (and therefore homozygotic) strains with particular characteristics are commercially available, as are many strains of knockout mice. Many human afflictions, including heart disease, diabetes, and cancer, can be modeled effectively in mice, and mice play a crucial role in developing preclinical data about potential human pharmaceuticals. The complexity of mouse biology makes them more difficult to manipulate and understand than many of the other model organisms, but also more like people. The mouse genome consists of 2,634,182,799 nucleotides in 20 diploid chromosomes and a mitochondrial chromosome. There are estimated to be more than 26,800 protein-coding genes. The numbers of alternatively spliced genes in the mouse are not known, but have been estimated to be as high as 60% of the total. The number of non-protein–coding RNA genes is known to be at least 3,000, but could be much higher. The model organism database for the mouse can be found at http://www.informatics.jax.org/.

10.3 Genetic Engineering

Advances in scientific knowledge often lead to advances in the engineering of useful entities and processes. The rapid strides in understanding of how life works have given people the ability to manipulate the genomes of living organisms by design, that is, to do genetic engineering.

Cloning is the word most powerfully associated with genetic engineering, but it is actually a relatively simple procedure with limited applications. Cloning refers to the process of making genetically identical offspring. The technological development that captured the popular imagination is the ability to take DNA from a somatic cell and use it to create an offspring with an identical genotype. Whereas the idea of producing an identical twin of an adult drives many horror stories and much moral concern, the routine use of the technique to create genetically identical laboratory animals for medical research has proven relatively uncontroversial, as well as useful.

The technology to add and remove genes from an organism has far more radical potential applications than cloning. The organisms produced in this way are called **transgenics**, since they include genes from some source other than their ancestors. Several steps are involved in creating transgenic organisms.

The starting point is creating a new DNA sequence that is to be inserted into the host organism. That sequence has to contain three types of information: (1) the protein coding sequence of the new gene to be inserted, (2) the regulatory sequences that will cause appropriate transcription factors to be recruited and the gene to be expressed at the desired time and place, and (3) other sequences that facilitate uptake of the gene into the host genome, called the **vector**, and allow for testing whether the gene was successfully transferred, called a marker. Collectively, the DNA sequence that is added to the host is called an insert.

Although it is possible to directly synthesize the DNA sequence of an insert, it is more common to produce it using **recombinant DNA**. Recombinant technology is a method for cutting DNA at specific sequences and then pasting two sequences together. The underpinnings of these methods are the highly specific restriction enzymes that bacteria use as an immune system (see section 8.3.2), along with DNA processing enzymes found in all cells. Libraries of useful sequences, such as protein-coding regions, transcription factor regulatory sites, and markers, are generally available, and can be straightforwardly assembled into an insert.

This **transgene** must then be incorporated into the DNA of the host. There are many methods that have been used to achieve this end, including the use of viruses. Viruses naturally use a variety of mechanisms to copy their own DNA into the host, several of which have been harnessed by bioengineers. However, the use of viral vectors has some drawbacks, including difficulty in targeting a specific place in the genome for the new DNA to be inserted.

Most vectors place inserts at random in the genome. One of the surprising failures of an early attempt at gene therapy (repairing defective genes that lead to human disease) was apparently due to disruptions of other genes by the insert.[7] Recently, methods have been developed that are better able to control the location in which an insert is placed; the essence of these approaches is to

---

7. The inability to control where an insert ended up in the genome was probably responsible for the failure of the attempts at human gene therapy for severe combined immunodeficiency. Though the defective gene was replaced, some of the inserts apparently led to cancerous transformations. See Seeking the cause of induced leukemias in X-SCID trial, *Science*, 24 (2003, Jan), 495.

detect and screen out cells where the insert happened at other than the desired location.

If reasonably long stretches (several hundred nucleotides) at the ends of the insert are the same as the sequence found in the host, homologous recombination (see section 5.2.3) will sometimes lead to the transgene replacing the host sequence. An embedded marker can be used to screen for this relatively rare event, and once the transgene has been properly incorporated into even a single host cell, many offspring can be produced by cloning. Homologous recombination is also the main technique used to produce knockouts. To produce a knockout, a damaged version of the target gene (say, with a premature stop codon) is associated with a marker used to screen for cells that took up the damaged gene.

Inducible knockouts, in which the expression of the gene is normal but can be shut down in some cell types or at some times under experimental control, require more complex constructs to be added to the genome. These constructs contain multiple genes with different regulatory regions. One of the inserted genes is a bacterial protein that cuts DNA at a particular sequence. The regulatory region associated with that gene is one that is recognized by a transcription factor present in the situation when the knockout is to be activated. The insert also replaces the original gene that is to be knocked out with a version that is bracketed by the sequence the bacterial protein recognizes. When the transcription factor activates the bacterial protein, the target gene is physically excised from the genome, resulting in an induced knockout.

As transgenic technology develops, the complexity of the inserts grows. As the example of the inducible knockout illustrates, it is not unusual for a system of several interacting genes to be inserted. As understanding of the interactions among naturally occurring gene products grows, it is likely that transgenics will be created with entire pathways or new molecular machines added.

Modification of genomes by human design is not the only important genetic technology. An approach known as directed evolution has also demonstrated impressive results recently. Since the domestication of animals and the dawn of agriculture, people have selected breeding stock to create plants and animals better suited to human needs. This venerable practice was the basis of Darwin's coinage of "natural selection," to describe the workings of evolution. Understanding of molecular biology has extended this idea into directed evolution of individual molecules.

Directed evolution of enzymatically active RNAs, called **selex**, has been the most successful of these new approaches. Selex has been effective at generating RNAs that bind to protein targets; the drug Macugen, for treating macular degeneration, is the first to become an approved pharmaceutical. Starting from a diverse pool of random RNA sequences, the RNAs that best bind to the intended target are chemically separated from the rest, and other sequences similar to those are created. The process is repeated until very high binding affinity sequences are discovered.

Although not yet practical, synthetic biology is an active field of research that proposes to create entire organisms by design.[8]

10.4   Molecular Biotechnology and Human Life

Although predictions of breakthroughs in genetic engineering have been made regularly since the 1970s, progress in the field has been uneven, and the business of genetic engineering has seen several cycles of boom and bust. Nevertheless, transgenic microorganisms now produce a large number of pharmaceuticals and other high-value substances, and genetically modified corn, wheat, and soybeans have had a substantial impact in agriculture. Engineered organisms able to produce industrial substances (e.g., liquid fuels) or remediate environmental problems (e.g., fix atmospheric carbon dioxide) are under commercial development.

The manipulation of the human genome for medical purposes, called **gene therapy**, has recently seen some modest successes in experimental applications. Transgenic manipulation of human beings is difficult, risky, and fraught with ethical concerns. However, the great potential of the idea and the imperative to help people with devastating or life-threatening conditions drive research forward. For example, the first successful use of gene therapy to treat cancer was reported in 2006. The otherwise untreatable malignant melanomas in two of seventeen patients regressed after infusion with the patient's own T-lymphocytes that had been engineered to express a receptor specific to their tumors.[9] Perhaps the most impressive success in gene therapy to date is the correction of the devastating (and often fatal) skin disease junctional epidermolysis bullosa by using a retrovirus to correct the causative genetic defect in skin stem cells that were removed from the patient, and then used to grow skin grafts, which remained stable and healthy.[10]

8. See http://syntheticbiology.org. The only comic ever to be published in the journal *Nature* was about synthetic biology: Adventures in synthetic biology, *Nature*, 438 (2005, Nov. 24), 449–453. Available as http://www.nature.com/nature/comics/syntheticbiologycomic.

9. Morgan et al. (2006), Cancer regression in patients after transfer of genetically engineered lymphocytes, *Science*, 314; 5796 (2006, Oct. 6), 126–129.

10. Fulvio Mavilio et al., Correction of junctional epidermolysis bullosa by transplantation of genetically modified epidermal stem cells, *Nature Medicine*, 12 (2006), 1397–1402.

Figure 10.8
Eduardo Kac's GFP bunny. Used with permission of the artist.

The era of genetic engineering is just beginning. Applications in medicine, agriculture, biofuels, and environmental remediation are clearly coming, although it not clear just how long they will take to arrive, nor in what form. It is perhaps fitting to end by noting two perhaps more surprising applications: Transgenic *Danio rerio* have become popular pets. Marketed as GloFish™, they are available in tropical fish stores for about $5, with a choice of green, yellow, and red fluorescent proteins. Perhaps even more striking is the work of Eduardo Koc, who calls himself a transgenic artist. One of his best-known creations is the GFP Bunny shown in figure 10.8; more of his art, and his writings that give it context can be found at http://www.ekac.org.

10.5   Suggested Readings

A large number of books cover various aspects of molecular biotechnology, although most require some comfort with mathematics and other technical background. Jurgen Gross's *Mass Spectrometry: A Textbook* (2nd ed., Springer, 2006) covers mass spec broadly, although it requires comfort with undergraduate math and physics. Although similarly technical, Gale Rhodes's *Crystallography Made Crystal Clear* (3rd ed., Academic Press, 2006) is probably about accessible as a book on protein crystallography can be. Deeper understanding of protein NMR requires some quantum mechanics, but for those up to it, John Cavenagh et al.'s *Protein NMR Spectroscopy: Principles and Practice* (2nd ed., Academic Press, 2006) is the standard textbook. Bernand Glick's *Molecular Biotechnology: Principles and Applications* (Cambridge University Press, 2006) is a broad and accessible book on recombinant DNA and genetic engineering techniques. Louis-Marie Houdebine's *Animal Transgenesis and Cloning* (Wiley, 2003) is similarly accessible, and focuses on genetic engineering of animals (rather than bacteria). *Stem Cell and Gene-Based Therapy: Frontiers in Regenerative Medicine* (Springer, 2007) is an edited volume that covers many of the most active areas of research. Evelyn Kelly's *Gene Therapy* (Greenwood Press, 2007) and *Stem Cells* (Greenwood Press, 2006) are part of the *Health and Medical Issues Today* series that tries to provide an even-handed background on the scientific, medical, legal, and ethical issues raised by these technologies. Suzanne Anker and Dorothy Nelkin's *The Molecular Gaze: Art in the Genetic Age* (Cold Spring Harbor Press, 2003) looks seriously at the relationship between biotechnology and art.

11 Molecular Bioethics

Having completed a brief tour of the major concepts of molecular biology, from evolution and chemistry to molecular medicine and biotechnology, it is now important to consider the meaning and significance of that material in a broader context, that is, to consider issues of bioethics. **Bioethics** is the study of value (what is good or bad) and morality (what is right or wrong) in biology and medicine. Application of molecular biological knowledge raises new ethical questions, and may offer some new approaches in grappling with older issues. Of course, judgments regarding bioethical issues depend on the broader viewpoints and social positions of the people making the judgments, and there are honest controversies about many of these questions. For that reason, the goal of this chapter is not to argue for any particular value proposition, but simply to survey some of the issues in which molecular biology is particularly important.

It takes a substantial period of time for the transformative power of new technologies to become apparent. Compare the current state of molecular biology to the development of the computer revolution. Though the basic theory underlying modern computation was laid out in the early part of the twentieth century, the first stored-program computer, ENIAC, was built in 1946. The first commercialization began shortly after, but was slow. A decade later, there were 112 computers in the world.[1] Half a century later, computers are ubiquitous, and have wrought fundamental changes in communication and finance as well as transforming industries as diverse as manufacturing, retailing, publishing, and music.

A reasonable date to set the origin of molecular biotechnology is 1974, with the first reported experimental production of recombinant DNA. That puts the biotechnology era at about half the age of the computer era. The steep growth

---

1. There were 66 IBM series 700 computers, and 46 UNIVACs, Cambell-Kelly & Aspray, *Computer: A history of the information machine*, p.127.

phase in the application of molecular biology is still ahead of us, although perhaps not by that much. The biggest impacts of computation have been in commerce and in communication. The impacts of biotechnology will likely be in more fundamental aspects of life, including what counts as disease, how disease is treated, what we eat, how we have babies—even what human capacities are and who is (or isn't) considered a human being.

11.1   Molecular Biology and Medicine

The first impacts of molecular biology in medicine are already becoming clear. Although the development of new treatments is generally seen as an unalloyed good, not all applications of molecular biology are as straightforward. For example, genetic testing, that is, testing for alleles that are indicative of the likelihood of developing a disease in the future, is fraught with complicated issues.

An increasing number of genetic tests are possible for diseases that do not have effective treatments. What use should be made of such tests? To some degree, getting such a test can be seen as a personal choice. Some people may want to know, and others prefer not to. However, such testing has broader implications. Should parents be allowed to test their children? Suppose that child, when grown, would rather not have known? Is it legitimate for other members of society to make use of the results (or the decision not to get a test), for example, in areas such as insurance or certification for dangerous or important professions?

Complex genetic predispositions, such as those related to heart disease or cancer, make issues in testing even more complex. With polygenic traits, it is not only the presence or absence of an allele that determines phenotype, but interactions among alleles and interactions between alleles and the environment. What should be done when genetic testing can offer incomplete, but potentially indicative information? Suppose genetic tests could indicate that certain environmental exposures were more (or less) dangerous for particular genotypes? How would our existing system of health and life insurance, environmental protection, and legal responsibility cope with such information?

The discovery of the first common allele related to a complex disease (PPAR gamma for diabetes) was published in 2000. In 2007, alleles from more than 50 genes were shown to change the risk of more than 20 common diseases, and it is likely that hundreds more such genes will be discovered shortly. Commercial organizations have started selling genotyping services, and the FDA is considering regulation. Many of the relationships that have been found are surprising. For example, one of the most clinically relevant of these is an allele of the gene TCF7L2 that nearly doubles the risk that prediabetics with the allele have of developing full-blown diabetes. The current main treatment for prediabetics is a drug called Metformin, which doesn't work as well for people with the high-risk allele of TCF7L2 as it does for the rest of the population. The good news, however, is that people with that allele respond much better to lifestyle interventions (diet and exercise) than people with the low-risk alleles. The high-risk TCF7L2 allele makes the environment a more important determinant of whether someone will get ill or not.

The promise of molecular biology in the future of medicine is sometimes summed up in the idea of personalized, predictive medicine. The idea is that each person's particular genotype could be used to predict likely health problems, and medicine (or lifestyle changes) could then intervene to prevent or ameliorate them. Of course, it is also possible that such predictions could be used to stigmatize or discriminate against individuals or groups. And as attractive as the idea of personalized medicine may be, it doesn't fit very well with the existing health care system in the United States and many other parts of the world.

A free market in health insurance may inevitably underinvest in prevention. If it takes more time for an investment in prevention to pay off with avoided health care costs is longer than people stay with one health insurance company, it doesn't make financial sense for insurers to pay for prevention. Predictive medicine doesn't work unless there is a mechanism to pay for the measures needed to avoid the predicted problems. State control doesn't necessarily result in adequate investment in prevention, either; many European national health care systems invest less in prevention than is spent in the United States.

One way to address this problem is to define the predisposing condition as an illness itself, even though it has no symptoms. Defining a high cholesterol level itself as a disease (now called hyperlipidemia) to be treated, rather than as just a risk factor for heart disease, opened the door to treatment reimbursement in the U.S. health care system. However, extending this approach to considering having a particular genotype as tantamount to having a disease has many other consequences with respect to how such people are treated (see following text). Are people with poor genetic prognoses but no symptoms disabled? How should they be insured?

Even the idea of personalized medicine doesn't fit well with the existing system. For example, the costs of developing new drugs are so high that commercial entities generally choose not to pursue areas without a large number of potential customers. Many drugs that might treat conditions that affect relatively small numbers of people languish because the potential return on investment is too low. The U.S. Orphan Drug Act attempted to address this issue, with somewhat mixed results. The idea of creating drug variants for particular genotypes makes the market for each individual drug even smaller. Changes in the ways drug development is done (and paid for) are likely to be necessary before personalized medicine becomes a reality.

One of the practices where advances in molecular biology have had the biggest impact to date is reproductive medicine. **In vitro fertilization** (IVF), in which egg and sperm are united outside of the body, has become widespread, and more than 200,000 Americans have been born by means of the technique since it was introduced in 1981. Genetic testing during IVF has given prospective parents unprecedented control over the genotypes of their children. Genetic testing is currently used to find disease markers, and even to have children who are genetically good matches for organ donations to ill siblings. Determining the sex of the embryo during IVF is also straightforward, and some institutions openly offer gender selection. IVF is rarely used by couples without severe infertility problems, but the procedure has been getting less expensive and less risky, a trend that is likely to continue, opening the door for people who simply wish to exercise more control over the genotype of their children.

Perhaps the most extreme form of control over the genotype of a child would be to create a clone, duplicating the genotype of another person. No human clones have been reported, and several countries have legislated against it. However, the process of creating a human clone is almost certainly no different from for other animals.

Gene therapy, the use of genetic engineering on human cells to cure disease discussed in section 10.4, remains experimental, although the first few successful applications are being reported (as have several deaths from experiments gone awry). It seems likely that the therapeutic transplantation of genes into somatic cells will eventually become part of medicine. A more difficult question is whether such manipulations should be allowed in the human germline. Such manipulations are likely to be technically feasible even today, although there is no evidence that they could be done safely. Eventually, replacement of a defective allele (say, for sickle cell disease) may be safer and more effective when done in the germline. It may be that germline therapy is the only way to genuinely cure some diseases. However, the prospects of making designed changed in the human germline quite reasonably raises important questions regarding the regulation, desirability, and potential social consequences of such changes.

Stem cells, particularly pluripotent embryonic stem cells, have shown a great deal of therapeutic potential, and have been effectively used to treat Parkinson's disease. However, some religious organizations object to this use of embryonic tissue, and the U.S. government has placed restrictions on it. Somatic cell nuclear transfer, also known as therapeutic cloning (where producing an offspring is not the goal), may be able to produce genetically matched pluripotent stem cells without destroying embryos.

Objections to the use of embryonic tissue arise because of disagreements about when an individual human life is thought to begin. Improved understanding of development and other advances in biotechnology have dramatically improved the care of prematurely delivered babies, steadily reducing the gestation required for viability. Since some definitions of the beginning of life depend on this idea of viability, important technologically driven change is likely to occur.

11.2 Molecular Biology and Agriculture

Food is perhaps even more central to human life than medicine. The use of genetically modified organisms (GMOs) in food has been widespread for some time now. Most Americans are not aware that a great deal of the food they eat is genetically engineered (most wheat-, corn-, and soy-based products, for example). One of the reasons that most consumers are unaware that GMOs are in their food is because the advantages of using the GMO accrue to agribusiness, rather than the consumer. Most of the genes that have been successfully added to food crops relate to the concerns of farmers and ranchers, such as yield and resistance to stresses such as pests, cold, drought, or the application of herbicides.

The very first attempt to market a GMO food was 1994's Flavr-savr tomato, which used RNAi (then called antisense technology) to suppress the expression of polygalacturonase, an enzyme that causes cell walls to soften during ripening. The idea was that these tomatoes could be picked ripe but still have a long shelf-life, and would therefore taste better. They were marketed for several years, but were a commercial failure. Consumer resistance to the idea of GMOs may have played a role in that failure, and still constrains the marketability of GMOs, particularly outside of the United States.

The concerns of those who oppose GMOs in food are primarily around safety. People who are allergic to certain substances may be exposed unexpectedly because of genetic engineering of a food in which the antigen would otherwise not be present. The accidental contamination in 2000 of the U.S. food supply with StarLink, a GMO corn approved only as animal feed, resulted in a successful class action lawsuit by people who claimed that they were harmed by allergic reactions. The StarLink transgene was a bacterial protein called Cry9C that prevented infestations by corn borers and cutworms.

Other concerns about use of GMO crops are about environmental consequences of using them. Perhaps the most striking of these is the concern about the flow of genes from one organism to another, in this context, called **genetic pollution**. The StarLink Cry9C gene was found in the corn of farmers who never planted the GMO, having gotten there by airborne pollen transmission or other routes. Another class action lawsuit in the StarLink case was won by farmers whose non-GMO crops had been contaminated; the damages awarded to the farmers were more than 10 times larger than that awarded to the allergy sufferers.

There have been other documented cases of transgenes jumping from one organism to another. In 2000, evidence was published suggesting that genes from GMO corn had been found in native stocks of corn in Oaxaca, Mexico. This flow of genes to landraces, the original varieties that were precursors of agricultural corn but that have themselves not been subject to modern breeding practices, is of particular concern. Landraces are used as genetic reservoirs in case of failures of agricultural crops, since the original varieties often harbor relevant genes that were lost during breeding. A governmental report in 2004 concluded that transgenes were likely present in Mexican landraces, although screening of a large sample in 2005 found no evidence of contamination.

Other environmental consequences of GMOs are also possible. Modified organisms may outcompete wild organisms, driving them to extinction. Aqua Bounty Technologies, Inc., has applied for FDA approval for a transgenic salmon that grows much more quickly than wild salmon. The modification was to the promoter region of a growth hormone gene. Normally, salmon turn on the gene only in the summer, but by replacing the promoter with one from another fish, the transgenic salmon grow year round. Concerns about the consequences of the fish escaping into the wild drove the company to seek approval only for fish that are to be grown indoors, although the company also claims that the fish would not be able to outcompete natural salmon in the wild.

11.3   Molecular Biology and Society

The concerns raised here about medicine and agriculture are primarily regarding the costs and benefits to individual patients, consumers, and farmers. There are also issues regarding the costs and benefits of this science (and resulting technologies) for society as a whole.

As in other scientific pursuits, molecular biologists are responsive to the priorities set by government funding agencies, charities that fund research, and commercial opportunities. The vast majority of scientific research and technical development in molecular biology is devoted to issues related to human health, with slowly increasing efforts in the areas of energy production and remediation of pollution. This work is undeniably important, but there are questions about whether research and development priorities properly reflect social needs. For example, there is little support for studies of molecular ecology, or related to ameliorating the consequences of mass extinction.

Another social concern is how the benefits of knowledge of molecular biology will be distributed, and what sort of property rights should exist around living things and their components. Although it is not possible to patent natural products, transgenic organisms, including mammals, have now been successfully patented, as have a wide variety of DNA-related inventions. More than 44,000 DNA-based patents have been issued.[2] Controversies about whether these patents encourage innovation or impede research, help or harm public health, and the effects they may have on global health disparities are ongoing. The World Health Organization published a useful assessment of these issues in 2006.[3]

Issues of global fairness also arise with regard to the use of indigenous knowledge of native plants in the development of molecular biotechnologies. An often-cited example is that of Madagascar's rosy periwinkle (*Catharanthus roseus*). Traditional healers used the plant, and as a result of natural products screens, two of the plant's compounds are now widely used cancer drugs (vinblastine and vincristine). Worldwide sales of the drugs exceed $150 million, but no compensation of any kind flowed to Madagascar.

Costa Rica, a small Central American nation, has become a global leader in defining the relationship between countries with potentially valuable organisms and molecular biotechnology industry. Although small, the country is rich in biodiversity, home to an estimated 4 to 5% of all terrestrial species. The government has protected about 25% of the area of the country from development, and created a nonprofit, public interest organization called Instituto Nacional de Biodiversidad (INBio) to exploit it. INBio has signed multimillion dollar agreements with Merck Pharmaceuticals to jointly research and develop drugs based on **bioprospecting** (screening natural products) in Costa Rica, and used those funds to develop a robust set of activities to protect biodiversity and use it to improve human life.

2. http://dnapatents.georgetown.edu/ is a publicly accessible database of all DNA patents and patent applications.

3. World Health Organization Commission on Intellectual Property Rights, Innovation and Public Health, published in the *Bulletin of the World Health Organization*, 84; 5 (2006, May), 337–424, available at http://www.who.int/bulletin/volumes/84/5/en/.

Concerns regarding the fairness of developments in biotechnology are not limited to the international sphere. The growing collection of human genotypes in the hands of the health care, the criminal justice (where human polymorphisms are used for identification), and other systems has led to concerns about unanticipated uses of this information for **genetic discrimination**. For example, if a particular genotype is associated with an increased risk for cancer or heart disease, can an insurance company deny coverage to a person who has it, or can an employer make a hiring decision based on that information? Some restrictions on genetic discrimination in the United States were enacted by President Clinton by executive order in 2000, and a bill is pending before the U.S. Congress as this is written that would substantially extend those protections.

A further concern is about discrimination not against individuals, but against particular groups. If alleles related to particular diseases (particularly mental illnesses or addiction disorders) were found to have a higher incidence in some populations than others, would that result in stigmatization of the group? The history of tests for the recessive traits that can cause sickle cell anemia (primarily found in African Americans) and those that can cause Tay-Sachs disease (primarily found in Ashkenazi Jews) give ample reason to believe that such test results can be misunderstood and misapplied to the detriment of the groups involved.

Other social concerns related to molecular biology are less immediate, but of profound significance despite their remoteness. The prospect of biowarfare or bioterrorism is particularly unsettling. Insights into molecular biology may make possible the production of particularly terrible weapons. As molecular technology improves, the cost and difficulty of producing such weapons is likely to fall dramatically. In 2002, a group of scientists synthesized an artificial polio virus, created entirely from purchased custom DNA sequences to demonstrate the reality of this concern. Vincent Racaniello, one of the scientists involved, said "Scientifically, the results are not surprising or astounding in any way. The point here, of course, is that the DNA can be synthesized from the sequence, and this could be done by any third-rate terrorist."

Perhaps the deepest social concern raised by molecular biotechnology is the question of what it means to be human. People have always used artificial means to improve their abilities; tools and cosmetics can both be seen as technologies for human enhancement. However, the use of molecular biotechnology for human enhancement opens previously unthinkable possibilities. As demonstrated by the doping controversies that engulfed both Tour de France (the world's greatest bicycle race) and the all-time home-run record set by baseball player Barry Bonds, athletic performance has already been forever changed by molecular enhancements. The dramatic expansion of use of botulism toxin (Botox) for cosmetic purposes, and the expansion of the tools of molecular biology into cosmetics is just the beginning of enhancing uses of these technologies.

People are highly motivated to improve their bodies and their abilities. As molecular biology produces insights into the mechanisms underlying human biological variability, and creates technologies that can manipulate those mechanisms, it seems inevitable that people will want to put that knowledge to work on improving themselves. It is entirely possible that these efforts will result in dramatic changes to the range of human phenotypes, and potentially call into question who is (or isn't) a human being.

11.4   Regulation and Control of Molecular Biotechnology

The dramatic possibilities raised by molecular biotechnology have created a strong social and political will to regulate the uses of this knowledge. Regulatory structures in the United States and around the world apply, as well as questions about whether additional mechanisms might be needed.

In the United States, federally funded experiments on human subjects are legally regulated by Institutional Review Boards (IRBs) to ensure that the rights and welfare of the subjects are protected. IRBs approve and monitor all experiments on people to ensure that they have scientific merit, are done with the informed consent of the subjects, and maximize safety.[4] Scientific research on vertebrate animals is similarly regulated by Institutional Animal Care and Use Committees (IACUCs) to ensure that it is scientifically or socially valuable, that the animals are properly cared for, that pain and suffering is minimized, and that only the minimum number of the most appropriate animals are used.[5] Privately funded experiments are much less constrained.

Commercial activities in molecular biology are regulated by a patchwork of agencies in the United States. The Food and Drug Administration (FDA) regulates food additives, cosmetics, and pharmaceuticals. The Environmental Protection Agency regulates pesticides and other toxins. The lack of clear boundaries can be confusing. For example, a plant with a pest-control

---

4. The U.S. law that regulates experimentation on human subjects is 45 CFR part 46, called The Common Rule. The Belmont Report, a key document in defining the ethics of experimentation on human beings, can be found at http://ohsr.od.nih.gov/guidelines/belmont.html. An article that describes some of the particular challenges with genetic experiments is Merz et al., Protecting subjects' interests in genetics research, *American Journal of Human Genetics*, 70; 4 (2002, Apr.), 965–971.

5. See http://www.absc.usgs.gov/research/vet/policies/IRACPRIN.htm for the regulations.

transgene, like the StarLink corn described previously, is regulated as a pesticide until it is harvested, and then it becomes a food.

Perhaps the largest controversy about the regulation of molecular biotechnology is around the application of what is called the **precautionary principle**. It has various versions, and its legal definition is complex, but the basic idea is that if there are credible scientific grounds for believing some new invention, technology, or product *might* not be safe, then it should not be introduced until there is convincing evidence about its risks and benefits. Whether this principle should be enshrined in government policy is probably the single biggest point of contention between the United States and Europe on biological issues. The Maastricht Treaty that created the European Union includes a version of the precautionary principle. The United States has claimed in the World Trade Organization proceedings that it is an illegal restraint of trade.

The opposition to the principle, annunciated most forcefully by the U.S. government, is that it is impossible to conclusively prove something new will be safe. The hypothetical risks that might be reasonable to consider are endless. The flow of innovation and new technology will be significantly impeded if every even hypothetical risk has to be addressed before the innovation can be adopted. Existing law already regulates demonstrated risks (such as toxicities or accidents) and those are the only ones that need to be considered. The European side argues that there have been a lot of examples in history when the consequences of new ways of doing things were not well understood when the practices began. Some of those innovations have turned out to have bad consequences that were difficult to remediate. It is much easier to prevent damage than it is to remediate it. So, when a society is evaluating tremendously potent technologies, it may make sense to impede innovation for a little while, to slow the process down, to avoid catastrophic mistakes.[6]

11.5 Learning about Life

It is not only the consequences of molecular biotechnology that are subject of moral and ethical concerns. Even simply teaching molecular biology, particularly any discussion of evolution, is repugnant to some communities. Both Christian and Muslim fundamentalists have actively opposed the teaching of

---

6. An enormous literature exists on the precautionary principle. One useful reference regarding its use around genetically modified organisms is H. van den Belt, Debating the precautionary principle: "Guilty until proven innocent" or "Innocent until proven guilty"? *Plant Physiology*, 132; 3 (2003, July), 1122–1126.

evolution, with some notable successes. However, history suggests that scientific knowledge spreads despite religious objections, and this book is a modest effort to ensure the basic ideas of molecular biology are accessible to all.

The existence of life is one of the most remarkable features of the universe, and its scientific understanding is one of the greatest achievements of humanity. Molecular biotechnology is changing the world, and may help address many of the greatest challenges facing humanity at the beginning of the twenty-first century. An understanding of molecular biology opens the door to many exciting opportunities: good careers, profitable businesses, and real contributions to the betterment of humanity can all be based on it.

Perhaps even more fundamentally, knowledge about life can be deeply satisfying in its own right. The ancient Greek aphorism to "know thyself" surely must include learning about the processes of being alive. Many find the molecular aspects of life as beautiful and inspiring as its more macroscopic manifestations.

Although this brief volume tries to introduce a lot of material, it is really only the barest beginning of the story. If you've made it this far, you have gotten over the activation barrier for knowledge of molecular biology, and you are now ready to take on whatever topics interest you, in as much detail as you care to. Go for it!

11.6 Suggested Readings

Bioethics is largely taken to mean biomedical ethics, and there are several excellent textbooks in that area: Robert Veatch's *The Basics of Bioethics* (Prentice Hall, 2002) is a gentle introduction, and Tom Beauchamp's *Principles of Biomedical Ethics* (Oxford University Press, 2001) is a more rigorous treatment, both focused more on clinical situations than molecular biotechnology.

Thousands of books are available on particular topics mentioned in this chapter, from hundreds of perspectives; too many to survey here. However, the President's Council on Bioethics publishes extensive free and often high-quality reports on ethical issues related to advances in biomedical science and technology, available from http://www.bioethics.gov. The National Institutes of Health offers an outstanding set of bioethics resources on the Web starting at http://bioethics.od.nih.gov, and the *American Journal of Bioethics* (http://www.bioethics.net) provides a wide-ranging forum for discussion of bioethical ideas.

Glossary

Bringing molecular biology into one's professional life requires becoming conversant with the terms used by its practitioners. Each technical term used in the book (even if it familiar from nontechnical uses) is defined here.

Recently, molecular biologists themselves have recognized the importance of carefully defining the terms they use to describe structures and functions of living things, and developed formal systems, called ontologies, to systematize names for the entities and processes they discuss. There is a National Center for Biomedical Ontologies (NCBO, http://bioontology.org) that provides a variety of services to the community. The Open Biomedical Ontologies (OBO, http://obofoundry.org) include freely redistributable ontologies that have been developed independently for different portions of biology; for example, the Sequence Ontology describes phenomena found in biomolecular sequences, and the Common Anatomical Reference Ontology provides anatomical terms that can be used to describe many organisms.

Whenever possible, the following terms are taken from an OBO ontology, as indicated in brackets at the end of each definition (some of which are exact matches, others of which map only to related terms in the ontology). The names of ontologies containing cross-referenced terms are abbreviated as follows:

ADW   Animal diversity web taxon database http://animaldiversity.ummz.umich.edu/site/about/technology/

CARO   Common anatomical reference ontology http://www.bioontology.org/wiki/index.php/CARO:Main_Page

CHEBI   Chemical entities of biological interest http://www.ebi.ac.uk/chebi/

CL   Cell type ontology http://obofoundry.org/cgi-bin/detail.cgi?cell

ENVO   Ontology of environmental features and habitats http://obofoundry.org/cgi-bin/detail.cgi?id=envo

EO   (Plant) Environmental conditions ontology http://obofoundry.org/cgi-bin/detail.cgi?id=plant_environment

FIX   Physicochemical methods and properties ontology http://obofoundry.org/cgi-bin/detail.cgi?fix

GO   Gene ontology http://www.geneontology.org/

PATO   Phenotypic quality ontology http://www.bioontology.org/wiki/index.php/PATO:Main_Page

REX   Physicochemical processes http://obofoundry.org/cgi-bin/detail.cgi?id=rex

SEP   Sample processing and separation techniques http://obofoundry.org/cgi-bin/detail.cgi?id=sep

SBO   Systems biology ontology http://www.ebi.ac.uk/sbo/

SO   Sequence Ontology http://www.sequenceontology.org/

TAXON   The National Library of Medicine taxonomy http://www.ncbi.nlm.nih.gov/Taxonomy/taxonomyhome.html/

All of these ontologies can be accessed through the NCBO's BioPortal (http://www.bioontology .org/bioportal.html).

3-phosphoglyceric acid   A 3-carbon compound that is an intermediate in the Calvin cycle. [CHEBI:3-phosphoglyceric acid]

3PGA   See *3-phosphoglyceric acid*.

3′ untranslated region   A sequence region at the 3′ end of a pre-mRNA that is not translated into a protein. [SO:three_prime_UTR]

5′ untranslated region   A sequence region at the 5′ end of a pre-mRNA that is not translated into a protein. [SO:five_prime_UTR]

Absolute fitness   The ratio of a genotype in a population before versus after selection. A ratio of greater than 1 means the genotype is being selected for; a ratio of 1 means the genotype is neutral; a ratio of less than one means the genotype is being selected against.

Acceptor   (electron) A substance to which an electron may be transferred. Often an atom with a valence number of 4 to 7 [GO:electron acceptor activity]

Acetyl-coenzymeA   An important intermediate molecule in several aspects of metabolism; a substrate in the tricarboxylic acid cycle. Also called acetyl-CoA. [CHEBI:acetyl-CoA]

Actin   A contractile protein found in muscle cells; with myosin, the molecular mechanism for muscle contraction.

Action potential   A rapid change in ion concentration (and hence voltage) that travels along a membrane; the mechanism by which signals are transmitted along neurons and cardiomyocytes. [GO:generation of action potential; GO:action potential propagation]

Activation   (of a molecule) A change in the structure of a molecule that realizes a functional effect in a cell.

Activation barrier   The amount of energy required for a reaction to pass through a transition state; see figure 3.6. [FIX:Gibbs free energy of activation]

Activation energy   The energy that is required for a reaction to take place. [FIX:Gibbs free energy of activation]

Active site   (of an enzyme) The region of an enzyme's three-dimensional structure where substrates are acted on. [FIX:protein active site residues; FIX:protein active site structure]

Active transport   The use of energy to transfer particular substances across a membrane. [GO: active transmembrane transporter activity]

Adaptive immunity   An immune response based on immunological memory, able to detect and respond to subsequent invasions by a previously encountered organism more quickly and effectively. [GO:adaptive immune response]

Adaptor protein   A protein that links an activated receptor protein with a particular signal transduction pathway, allowing for cellular response to be based on complex combinations of signals. [GO:transmembrane receptor protein tyrosine kinase adaptor protein activity]

Adenine   One of the four nucleotide bases found in DNA, pairs with thymine; also a component of the energy carrying molecules nicotinamide adenine dinucleotide and adenosine triphosphate. [CHEBI:adenine]

Adenosine triphosphate   The main energy-carrying molecule of cells; also called ATP. [CHEBI: ATP]

Advantageous mutation   A mutation that increases the fitness of an organism.

Agonist   A compound that interacts with a receptor to increase the proportion of receptors in the activated form. [GO:receptor agonist activity]

Alanine   An amino acid; see table 5.1 for its properties. [CHEBI:alanine; CHEBI:alanine residue]

Allele   One of several alternate forms of a gene. A single allele is inherited from each parent. [SO:allele]

Allometry   Change in the relative sizes of different body parts, leading to a significant change in the form of an organism.

Allosteric regulation   (of an enzyme) Regulation of the activity of a protein through a molecular interaction at a position other than that of the protein's active site. [SBO:allosteric control]

Alpha carbon   (of an amino acid) The central carbon atom in an amino acid. In organic chemistry, carbon atoms are named in order by Greek letters (alpha, beta, gamma, etc.) starting with the carbon attached to the main functional group.

Alpha helix   (plural alpha helices) A protein secondary structure that has a compact, corkscrew shape. [SO:alpha_helix]

Alpha-ketoglutarate   A compound in the TCA cycle that is also important as one of the routes by which nitrogen is brought into metabolism (producing glutamate). [CHEBI:2-oxoglutarate(2-)]

Alternative splicing   The processing of an RNA transcript into multiple different mRNA molecules by including some exons and excluding others. [GO:alternative nuclear mRNA splicing, via spliceosome]

Amino acid   An organic acid containing an amine (-NH₂) group; the monomeric building block of polypeptides and proteins. [CHEBI:amino acids; CHEBI:amino-acid residues]

Amino acid polymerase   An enzyme that extends a polypeptide by catalyzing the addition of another amino acid to it.

Amino acid sequence   The sequence of amino acids that makes up a protein or polypeptide. Also called the primary structure of a protein. [FIX:amino acid sequence]

Amino group   A chemical functional group containing one nitrogen and two hydrogen atoms, resembling ammonia. [CHEBI:amino group]

Aminoacyl-tRNA synthetase   A multienzyme complex that catalyzes the attachment of an appropriate amino acid to its corresponding tRNA. [GO:aminoacyl-tRNA synthetase multienzyme complex]

AMPA receptor   A protein complex found on the surface of neurons that is activated by the neurotransmitter (and amino acid) glutamate, causing inflows of calcium into the postsynaptic cell. [GO:alpha-amino-3-hydroxy-5-methyl-4-isoxazoleproprionic acid selective glutamate receptor complex]

Anabolism   A metabolic process that synthesizes complex molecules out of simpler constituents. [GO:biosynthetic process]

Anaphase   The stage in mitosis during which the chromosomes begin to separate. [GO:anaphase]

Anatomy   The study of the structure of organisms.

Angina   Chest pain caused by reduced flow of blood to the heart. [DOID:Intermediate coronary syndrome]

Anion   An ion with a negative charge. [CHEBI:anions]

Antagonist   A compound that interacts with a receptor to decrease the proportion of receptors in the activated form. [GO:receptor antagonist activity]

Antenna complex   (in photosynthesis) Proteins in the thylakoid membrane that transfer light energy to a chlorophyll molecule in a photosystem. [GO:photosystem I antenna complex; GO:photosystem II antenna complex; GO:B800-820 antenna complex; GO:B800-850 antenna complex; GO:B875 antenna complex]

Anterior-posterior axis   An imaginary line running from the head (anterior) to the tail (posterior) of an organism [GO:anterior/posterior axis specification; PATO:anterior to; PATO:posterior to]

Antibody   A protein produced by the immune system to recognize and neutralize foreign molecules; also called an immunoglobin. [GO:antigen binding; FMA:Immunoglobulin]

Antigen   A molecule or substance that triggers an immune response.

Antimicrobial peptides   Short proteins used by many organisms to attack bacteria; part of the innate immune system.

Antiparallel beta sheet   A protein secondary structure that consists of two beta strands running in opposite directions that are hydrogen bonded to each other. [FIX:antiparallel beta-sheet]

Apolipoprotein   Lipid-binding proteins that package lipids for aqueous transport, particularly through the blood. [CHEBI:apolipoproteins]

Apoptosis   A normal cellular process that leads to self-destruction and death of a cell; one way a multicellular organism gets rid of unnecessary or damaged cells. [GO:apoptosis]

Aqueous   Related to water. An aqueous solution is one in which the solute is water.

Archaea   A group of single-celled organisms; one of the three major divisions in the most widely accepted classification of organisms. [TAXON:Archaea]

Arrest   (in cell cycle) Halting the process of cellular division, often in response to damage. [GO: cell cycle arrest]

Arteriole   A small diameter blood vessel that connects arteries to capillaries. [FMA:Arteriole]

Artery   A vessel that carries blood away from the heart. [FMA:Artery]

Assortative mating   Nonrandom sexual reproduction on the basis of phenotype; i.e., mating of individuals that are more similar to (or more different from) each other than would be expected by chance.

Atherosclerosis   Narrowing of arteries due to buildup of fatty deposits on their inner linings. [DOID:Atherosclerosis]

Atom   The smallest amount of an element that retains the chemical properties of that element. [CHEBI:atoms]

Atomic number   The number of protons in the nucleus of an atom. [FIX:atomic number]

Atomic mass   The abundance-weighted average of the masses of an element's isotopes; roughly the sum of the number of protons and neutrons in an element.

ATP   See *Adenosine triphosphate.*

ATP synthase   A class of enzymes that uses the flow of protons across a membrane to synthesize ATP from adenosine diphosphate and inorganic phosphate. [GO:proton-transporting ATP synthase complex]

Atrium   (plural atria) One of the upper two chambers of the mammalian heart, where blood is collected as it returns from the body. [FMA:atrium]

Atrophy   The shrinkage of body part as a result of a reduction in the size or number of cells that constitute it. [PATO:atrophied]

Autophagy   A process in which cells digest parts of their own cytoplasm. [GO:autophagy]

Autophosphorylate   The addition of a phosphate group to a protein kinase through its own enzymatic activity. [GO:protein amino acid autophosphorylation]

Autotroph   An organism that requires only inorganic materials to live, and does not require the products of any other organisms; carbon dioxide is its source of carbon. [GO:carbon utilization by fixation of carbon dioxide]

Axon   The part of a nerve cell, often a long, branching extension, that transmits outgoing signals. [GO:axon]

B-cell   The type of white blood cell that produces antibodies. Also called a B-lymphocyte. [CL: B cell]

Backbone   (of a polypeptide) The alpha carbons, carboxyl carbons, amino nitrogens, and the bonds that link them in a polypeptide or protein.

Balancing selection   A kind of natural selection that tends to maintain multiple alleles within a population.

Baroreceptor   A sensory neuron that responds to pressure, particularly blood pressure. [FMA: Baroreceptor; GO:baroreceptor feedback regulation of blood pressure]

Base pair   (of DNA) Two complementary nucleotides held together by hydrogen bonds; the unit of measure for the size of a double-stranded nucleic acid sequence. [SO:base_pair]

Basement membrane   Specialized layers of extracellular matrix that separate epithelial tissue from underlying connective tissue. [FMA:Basement membrane]

Bases   (in nucleotides) The portion of a nucleic acid that is involved in pairing. A base bound to a ribose or deoxyribose sugar is called a nucleoside; a nucleoside with one or more phosphate groups attached is called a nucleotide. [CHEBI:nucleobases]

Benign   A medical condition or anatomical malformation that is not dangerous to health; often used to describe tumors that are not cancerous. [DOID:Benign Neoplasm]

Beta cell   A type of cell found in the pancreas that produces insulin. [CL:pancreatic B cell]

Beta oxidation   A metabolic process that consumes fatty acids and produces acetyl-CoA, which is then fed into the TCA pathway to produce ATP. [GO:fatty acid beta-oxidation]

Beta sheet   A protein secondary structure that consists of two beta strands that are hydrogen bonded to each other. [FIX: antiparallel beta-sheet; FIX:parallel beta-sheet]

Beta strand   A protein secondary structure that has an extended, rodlike shape. [SO: beta_strand]

Bilateria   The group of animals that exhibit bilateral symmetry during some portion of their lifespan. [TAXON:Bilateria]

Binary fission   (of cells) Nonsexual cell division that results in offspring of approximately equal size. [GO:binary fission]

Binding affinity   The strength of attraction between two molecules, usually a protein and a ligand. [ECO:inferred from ligand binding]

Bioethics   The study of moral values and obligations in biology and medicine.

Biomass   The total mass of living organisms in an area or community.

Bioprospecting   The exploration of the organisms in an ecosystem for commercial, scientific, or culturally valuable genetic or biochemical resources.

Blastula   An early stage of embryonic development in animals beginning when the cells of a morula rearrange to form a sphere. [FMA:blastocyst]

Bond energy   The energy required for (or produced by) the breakage of a chemical bond; a measure of the strength of a chemical bond.

Bone morphogenetic proteins   A family of protein growth factors that drive the differentiation of cells into bones, cartilage, teeth, and similar tissues.

Brain   An organ containing most of the interneurons that relate sensations to actions; with the spinal cord, forms the central nervous system. [FMA:Brain]

Budding   (in mitosis) Nonsexual reproduction that involves the development of an offspring from a small outgrowth of the parent. [GO:cell budding]

C terminus   (of a polypeptide) The end of a polypeptide or protein that is terminated by a free carboxyl group; also called the carboxy terminus; conventionally the end of a protein, since translation proceeds from N terminus to C terminus. [CHEBI:C-terminal amino-acid residues]

Cadherins   A family of transmembrane proteins that play a key role in the differential adhesion of cells to one another; they are cell-type specific and depend on calcium ions to function. [FMA: Cadherin]

Calcium-dependent kinases   A family of signal transduction molecules that respond to increased intracellular calcium levels by phosphorylating other proteins; these proteins are found primarily in neurons, and play an important role in altering neuronal connection strength as a result of experience. [GO:calcium- and calmodulin-dependent protein kinase complex]

Calvin cycle   A metabolic pathway in plants that produces sugars and other complex carbohydrates from carbon dioxide. [GO:reductive pentose-phosphate cycle]

cAMP   See *Cyclic adenosine monophosphate*. [CHEBI:3′,5′-cyclic AMP]

Capillary   The smallest type of blood vessels in an animal; exchange of nutrients and wastes between the blood and the interstitial fluid takes place through capillary walls. [FMA: Capillary]

Carbohydrates   (also called saccharides) Organic compounds containing only carbon, hydrogen, and oxygen; these include sugars and starches. [CHEBI:carbohydrates]

Carboxyl group   An acidic chemical functional group that contains a carbon atom double bonded to an oxygen atom and single bonded to another oxygen atom, which in turn is bonded to a hydrogen atom. [CHEBI:carboxylato group]

Cardiomyocyte   A heart muscle cell. [CL:cardiac muscle cell]

Caspases   A family of proteases that play a central role in apoptosis. [GO:caspase complex; GO:caspase activity]

Catabolism   A metabolic process that breaks down large molecules into smaller ones, generally releasing energy. [GO:catabolic process]

Catalysis   Promoting or increasing the rate of a chemical reaction with a catalyst, a substance that is left unchanged by the reaction. [GP:catalytic activity]

Cation   A positively charged ion. [CHEBI:cations]

cDNA   See *Complementary DNA*. [SO:cDNA_clone]

Cell   The simplest structure that supports independent life; the basic structural and functional unit of life. [CL:cell; GO:cell]

Cell adhesion molecules   A family of transmembrane proteins that play a role in the binding of a cell to other cells or to the extracellular matrix. [GO:cell adhesion; FMA:Cell adhesion molecule]

Cell cycle   A repeating sequence of events that eukaryotic cells undergo resulting in mitosis. [GO:cell cycle]

Cell fate   The ultimate differentiated or specialized state that a cell in a multicellular organism will become; cell fate determination precedes differentiation. [GO:cell fate commitment; GO:cell fate determination; GO:cell fate specification]

Cell fate commitment   The process by which the fate of a cell is set. [GO:cell fate commitment]

Cell-mediated immune response   The cytotoxic response to foreign cells; the portion of adaptive immune system that does not involve antibodies. [GO:cell activation during immune response]

Cell membrane   A lipid bilayer that forms the outer boundary of a cell, separating it from its environment. [GO:plasma membrane]

Cell type   A distinct morphological or functional form of a cell. [CL:cell; GO:cell]

Cellular respiration   The catabolic processes that produce energy through the use of oxygen. [GO:oxidative phosphorylation]

Central dogma   Watson and Crick's central claims about molecular biology: that information about how to make proteins was encoded in DNA, that DNA can be copied to make more DNA or transcribed into RNA, which codes for proteins. In essence, information flow is always DNA to DNA or DNA to RNA to protein.

Centrifuge   An instrument that rapidly spins solutions to separate their solutes by density. [FIX: centrifugation]

Centromere   The central portion of a diploid chromosome where the two chromatids are joined. [GO:chromosome, pericentric region]

Cerebellum   A large portion of the brain, generally found toward the base of the skull; thought to be responsible for long or complex sequences, including walking and balance. [FMA: Cerebellum]

Cerebral ganglion   A mass of interneurons that plays the role of a brain in many insects.

Channel   (in a membrane) A microscopic opening in a biological membrane that can be selectively opened and closed. [GO:channel activity]

Checkpoint   (in cell cycle) Specific points during the process of cell division where the integrity of the DNA or other critical factors are verified before division can continue. [GO:cell cycle checkpoint]

Chemical fossil   Remnants of organisms in the form of compounds or isotope ratios that could only have been produced by living things.

Chemotaxis   The motion of a cell or organism in response to a chemical stimulus. [GO: chemotaxis]

Chip-chip   An experimental procedure that provides information about all of the positions in a genome that are bound by a particular transcription factor.

Chiral   Not having a plane of symmetry; a molecule that has distinct left- and right- handed forms.

Chlorophyll   A family of compounds that absorbs light and is central in the process of capturing energy through photosynthesis. [CHEBI:chlorophylls]

Chloroplast   An organelle present in algae and plants where photosynthesis takes place. [GO: chloroplast]

Chromatid   One of the two strands of chromatin that makes up a diploid chromosome.

Chromatin   The complex of DNA and proteins that is the structure underlying chromosomes. [GO:chromatin]

Chromosome   The part of a cell that embodies its genotype. [GO:chromosome]

Cilia   Short, hairlike structures attached to the outside of a cell membrane. [GO:cilium]

Circulatory system   The portions of a multicellular body through which liquids (usually blood, but also lymph) are transported. [FMA:Cardiovascular system]

Cis   Latin for "on the same side" or "nearby." For a gene, it means something in the DNA near the position of the gene.

Cis regulatory sequence   A segment of DNA, usually near a gene, which is recognized and bound by regulatory proteins that influence the expression of the gene. Also called a regulatory sequence. [SO:transcriptional_cis_regulatory_region]

Citrate   An ionic form of citric acid that is one of the substrates in the TCA cycle. [CHEBI: citrate(3-)]

Citric acid cycle   See *Tricarboxylic acid cycle.*

Clonal colony   A population of genetically identical organisms (usually bacteria).

Clone   An organism with precisely the same genotype as another. [SO:clone]

Cloning   Producing an organism with precisely the same genotype as another. Also sometimes used to mean producing a copy of an individual gene by recombination into a vector. [SO: clone]

Coagulation cascade   The set of chemical reactions that leads to blood clotting. [GO:coagulation]

Coding sequence   A DNA sequence that specifies the sequence of a protein or functional RNA. [SO:CDS]

Codon   A sequence of three nucleotides that specifies an amino acid. [SO:codon]

Coelom   The fluid-filled cavity in the body of an animal, lined by epithelium derived from mesodermal cells. [FMA:Extraembryonic celom]

Coenzyme   An organic molecule, usually derived from a water-soluble vitamin, that combines with and activates specific enzymes.

Coevolution   The evolution of two or more species whose characteristics mutually influence each other's fitness; for example, predators and prey, or plants and their pollinators.

Collagen   A fibrous protein that is the main structural component of connective tissue. [GO: collagen; FMA:Collagen]

Colony   A population of organisms of the same species living closely together, often to some mutual benefit.

Combinatorial chemistry   The rapid synthesis of a large number of distinct compounds by forming combinations of an initial set of chemical constituents, often used to create libraries of potential pharmaceuticals for screening.

Commensal   An organism that lives in close association with and derives benefit from another organism, without either harming or helping the other. [GO:symbiosis, encompassing mutualism through parasitism]

Complement system   A part of the innate immune system consisting of a set of proteins that attack the membranes of foreign cells. [GO:membrane attack complex; GO:activation of membrane attack complex; GO:complement activation]

Complementary DNA   A DNA molecule synthesized in a laboratory from a messenger RNA template using the enzyme reverse transcriptase; it contains only the nucleotides of a coding sequence. [SO:cDNA_clone]

Complementary base   A nucleotide base that forms hydrogen bonds with another; for example, adenine is the complementary base to thymine.

Complementary strand   (of DNA) A strand of DNA is complementary to another strand if it is made of the complementary bases in the opposite order. Two complementary strands of DNA will hybridize to form a double-stranded DNA molecule.

Compound   A pure substance that cannot be separated into components by physical means; equivalently, a substance formed by the chemical combination of two or more elements in a definite proportion. [SEP:substance; CHEBI:polyatomic entities]

Conformation   The shape or particular three-dimensional structure of a molecule or other entity. [SO:structural_region; FIX:conformation]

Connective tissue   One of the four basic types of tissues found in animals; it creates the structural components that hold organs together and in place; it also produces the extracellular matrix. [FMA:Connective tissue]

Coprolites   Fossilized feces.

Core   (of a protein) The tightly packed central region of a protein.

Core metabolism   The metabolic activities that are similar in most organisms. [GO:primary metabolic process]

Covalent bond   A chemical bond in which electrons are shared among two or more atoms. [FIX: covalent bond]

Crossovers   The points in meiotic cell division where there is a change in which parental chromosome is the source of genetic material for an offspring. [GO:meiotic recombination]

Cyclic   (compound) An organic molecule that contains a series of carbon atoms bonded together into a loop or ring. [CHEBI:cyclic compounds]

Cyclic adenosine monophosphate   A cyclic compound that often acts as a signal within or between cells; chemically, a relative of the nucleotide adenosine and the energy carrier ATP. [CHEBI:3′,5′-cyclic AMP]

Cysteine   A sulfur-containing amino acid; see table 5.1 for its properties. [CHEBI:cysteine; CHEBI:cysteine residue]

Cytochrome b6f complex   A group of proteins in the electron transport chain in photosynthetic organisms. [GO:cytochrome b6f complex assembly]

Cytokine   A glycoprotein secreted by immune cells that signals other immune cells and often has some cytotoxic effect itself.

Cytology   The study of cellular structure.

Cytosol   The internal fluid of a cell.

Cytoplasm   The complex mixture of proteins and other compounds that supports metabolism; all of the contents of a eukaryotic cell that are not the nucleus or cell membrane. [GO: cytoplasm]

Cytosine   A nucleotide base found in nucleic acids; pairs with guanine. [CHEBI:cytosine]

Cytoskeleton   A three-dimensional network of protein structures in the cytoplasm that provides internal support, holds cellular structures in place, and plays a role in cell shape and movement. [GO:cytoskeleton]

Cytotoxic   Capable of killing cells; usually a reference to the types of immune cells that can kill other cells. [CL:CD8-positive, alpha-beta cytotoxic T-cell]

Dehydrogenase   The class of enzymes that catalyze the transfer of hydrogen atoms; usually this results in the oxidation of a substrate, and the transfer of electrons to an acceptor.

Deleterious mutation   A change to the genotype of an organism that results in reduced fitness.

Deletion   (of DNA) The loss of one or more nucleotides from a DNA sequence. [SO: nucleotide_deletion]

Denature   To cause a molecule to change structure from its normally occurring form, usually resulting in the loss of its biological function. [GO:protein denaturation]

Dendrite   The part of a nerve cell, often a long, branching extension, that receives incoming signals [GO:dendrite; FMA:Dendrite]

Dendritic tree   The entire collection of dendrites associated with a single nerve cell.

Deoxyribonucleic acids   Linear polymers of deoxyribonucleotides whose biological function is to encode information; also called DNA. [CHEBI:deoxyribonucleic acids]

Deoxyribonucleotide   A dinitrogen heterocycle bonded to deoxyribose and a phosphate group; the monomeric unit of deoxyribonucleic acids (DNA).

Deoxyribose   A five-carbon sugar (ribose) with one oxygen atom removed (deoxy-) [CHEBI:2-deoxy-D-ribose]

Development   The process of creating a complete multicellular organism from a single cell. Also sometimes used to refer to a specific portion of that process, e.g., the development of an organ. [GO:developmental process]

Diastole   The period of time when the heart ventricles dilate or expand; the interval between ventricular contractions.

Differentiation   The process by which a cell or tissue acquires its specialized characteristics and functions. [GO:cell differentiation]

Digestive system   The group of organs that collectively ingest, break down and absorb nutrients. [FMA:Alimentary system]

Dihedral angle   The angle formed by two intersecting planes.

Dimer   A compound formed by the union of two other compounds; often a complex of two proteins. [GO:protein dimerization activity]

Dimerize   To form a dimer. [GO:protein dimerization activity]

Dinitrogen heterocycle   A cyclic compound in which the ring of atoms includes both carbon atoms and two nitrogen atoms. Ribose and deoxyribose are dinitrogen heterocycles.

Diploid   Having two copies of each chromosome (and therefore two alleles for each gene). Contrast with haploid or polyploid. [PATO:diploid]

Directional selection   A process of natural selection that favors phenotypes at an extreme of some range.

Disease   An abnormal condition that impairs biological function. [DOID:Disease]

Dissociation   (reaction) The separation of a compound in aqueous solution into ions. [REX: dissociation]

Distal   (See *Proximal-distal axis*) Farther away, usually from the center of the body [PATO:distal to]

Diversifying selection   See *Balancing selection.*

DNA   See *Deoxyribonucleic acids.*

DNA-directed DNA polymerase   An enzyme or enzyme complex that catalyzes the creation of a new DNA molecule that has a complementary sequence to an existing DNA molecule. [GO: DNA-directed DNA polymerase activity]

DNA-directed RNA polymerase   An enzyme or enzyme complex that catalyzes the creation of a new RNA molecule that has a complementary sequence to an existing DNA molecule. [GO: DNA-directed RNA polymerase complex]

DNA ligase   An enzyme that catalyzes the formation of a phosphodiester bond between adjacent but unbonded nucleotides, linking two strands of DNA. [GO:DNA ligase activity]

DNA polymerase III   The enzyme complex that plays a central role DNA replication in bacterial cells. [GO:DNA polymerase III complex]

DNA primase   A DNA-directed RNA polymerase that catalyzes the formation of short RNA primers during DNA replication. [GO:DNA primase activity]

DNA repair mechanism   A process by which a cell identifies and corrects damage to DNA molecules. [GO:DNA repair]

DNA sequencer   An instrument that determines the nucleotide bases (in order) that constitute a DNA molecule. [FIX:DNA sequencing]

Dominant   (allele) An allele that produces the same phenotype in heterozygous and homozygous organisms.

Donor   (electron) An atom that tends to lose one or more electrons in chemical reactions. [GO: electron donor activity]

Dorsal-ventral axis   An imaginary line running from the belly (ventral) to the back (dorsal) of an organism. [GO:dorsal/ventral axis specification; PATO:dorsal to; PATO:ventral to]

Double bond   A covalent bond where two pairs of electrons are shared between atoms. [FIX: double bond]

Double-stranded DNA   A molecule made of two complementary strands of DNA held together by hydrogen bonds; the form in which DNA molecules are usually found in most organisms. [CHEBI:double-stranded DNA]

Double-stranded RNA   A molecule made of two complementary strands of RNA held together by hydrogen bonds; an unusual form of RNA molecule that plays a role in the process of RNA interference.

dsRNA   See *Double-stranded RNA.*

ECM   See *Extracellular matrix.*

Ecosystem   An interrelated community of diverse organisms and their physical and chemical environment. [EO:abiotic environment; EO:biotic environment]

Ectoderm   The outermost layer of the gastrula, and all of the cells that descend from it. [FMA: Ectoderm; GO:ectoderm development]

Electrolytes   Ions present in the blood and interstitial fluid. [GO:ion homeostasis]

Electron   A negatively charged subatomic particle responsible for many chemical phenomena. [CHEBI:electron]

Electron transport chain   A spatially separated series of chemical reactions that gradually tap the energy in high-energy electron donors to create a proton gradient across a membrane, ultimately producing ATP. [GO:electron transport; GO:electron transporter activity]

Electronegativity   A measure of the ability of an atom to attract electrons to itself in a covalent bond.

Electronic structure   The physical distribution of and relationships among the electrons in a compound.

Electrostatic force   The force between two charged particles; like charges repel, unlike charges attract.

Elements   The set of compounds that cannot be separated into simpler substances by chemical means.

Elongation factors   (for ribosomal protein synthesis) The enzymes that catalyze the addition of amino acids to the end of a growing polypeptide chain at a ribosome. [GO:translation elongation factor activity]

Emergent property   A property of an assembly of components that depends on the relationships among those components; often one that is considered surprising based on the properties of the individual components.

Endocardium   The endothelial lining of the heart chambers and valves. [FMA:Endocardium]

Endocrine system   A collection of ductless glands that regulate many body functions through hormonal secretions. [FMA:Endocrine system]

Endoderm   The innermost layer of the gastrula, and all of the cells that descend from it. [FMA: Endoderm; GO:endoderm development]

Endonuclease   An enzyme that cleaves the phosphodiester bonds joining nucleotides within a nucleic acid; it often recognizes and cuts at only a particular nucleotide sequence. [GO: endonuclease activity]

Endoplasmic reticulum   A cellular organelle consisting of a membrane network of tubes and sacks that is involved in the production and transportation of proteins. [GO:endoplasmic reticulum]

Endosymbiosis   The close association of two organisms, one of which lives inside the other.

Endothermic reaction   A chemical reaction in which heat is absorbed from the surroundings. [REX:endothermic reaction]

Energy   The capacity of a system to do work. [PATO:energy]

Enhancer   (of gene expression) A cis regulatory DNA sequence that tends to increase the expression of a coding sequence; usually meaning one that is not the coding sequence's promoter. [SO:enhancer]

Enthalpy   In thermodynamics, the heat content of a system, denoted H; it cannot be measured directly, but changes, denoted ΔH, can be measured.

Entropy   The amount of energy in a system that *cannot* be used to do work.

Environment   The sum of all external conditions and influences that affect an organism or group of organisms. [EO:abiotic environment]

Enzyme   A biological molecule (usually a protein, but sometimes also an RNA) that acts as a catalyst. [SBO:enzyme]

Epithelium   A type of tissue that covers a surface or lines a cavity in a multicellular organism; generally epithelium functions to protect, secrete, or absorb. [CARO:epithelium; FMA: Epithelium]

Equilibrium   A stable state in which opposing forces or fluxes balance each other; in a chemical reaction, when the forward and backward reactions proceed at equal rates, leaving the concentrations of the reactants and products unchanged.

Equilibrium constant   In a chemical reaction, the ratio between the concentration of the products and the concentration of the reactants when the reaction is in equilibrium. [SBO:equilibrium constant]

ER   See *Endoplasmic reticulum.*

Erythroblast   A bone marrow cell with a nucleus from which red blood cells develop. [CL: erythroblast]

Erythrocyte   See *Red blood cell.*

Essential genes   Genes that are required for life; in multicellular organisms, only genes that are required to reach a live birth are generally considered essential.

Essential oils   Mixtures or compounds extracted from plants or plant parts and used by people for therapeutic or cosmetic benefits. [CHEBI:volatile oils]

Etiology   The causes or origins of a disease.

Euchromatin   Chromosome regions that are loosely packaged and relatively accessible to RNA polymerases. Contrast with heterochromatin. [GO:euchromatin]

Eukaryote   A group of organisms composed of cells that contain a membrane-enclosed nucleus; one of the three major divisions in the most widely accepted classification of organisms. All multicellular organisms are eukaryotes. [TAXON:Eukaryota]

Evo devo   See *Evolutionary developmental biology.*

Evolution A process of change that involves reproduction with inherited characteristics, a source of variation in those inherited characteristics, and differential reproductive success depending at least in part on the inherited characteristics.

Evolutionary developmental biology The study of the evolution and function of the genes and gene products that influence development from egg to mature adult.

Excitable cells Cells that can be triggered to change based on electrical potential; nerve cells and muscle cells are both excitable.

Excitatory neurotransmitter A substance released by one nerve cell and detected by another that acts to elicit an action potential or make it more likely that one will be elicited.

Exon A segment of DNA that contains the information for part or all of a gene product. [SO: exon]

Exonuclease An enzyme that catalyzes the cleavage of nucleotides from the end of a nucleic acid. [GO:exonuclease activity]

Exothermic reaction A chemical reaction in which heat is released into the surroundings. [REX: exothermic reaction]

Expression (of a gene) The creation of a gene product (protein or functional mRNA) from a gene. [GO:transcription; GO:translation]

Expression (of an allele) The manifestation of an allele in the phenotype of an organism.

Expression array An instrument that assays the expression of a large number of genes simultaneously, through hybridization of mRNAs. [ECO:inferred from expression microarray experiment]

Extracellular Outside of a cell. [GO:extracellular space]

Extracellular matrix Any material produced by cells and secreted into the surrounding medium, but usually the mixture of fibrous elements, proteins, and polysaccharides secreted by connective tissue. [GO:extracellular matrix]

Fascicle A bundle of muscle fibers surrounded by connective tissue. [FMA:Fascicle]

Fatty acid A molecule that consists of a lipid chain terminated by a carboxyl group. [CHEBI: fatty acids]

Fibroblast A type of connective tissue cell that secretes collagen and other extracellular matrix proteins; plays an important role in wound healing. [CL:fibroblast]

Fibronectin A family of extracellular matrix proteins that bind cell surface proteins to other extracelluar matrix proteins, providing support for tissues and organs; also plays a role in wound healing. [FMA:Fibronectin]

Fitness Contribution of a trait to reproductive success.

Fitness landscape A graphical representation of the relationship between the reproductive success, the environment, and the phenotype of an organism.

Fixation (of a trait) The propagation of a trait throughout a population.

Flagellum (plural flagella) A long, thin structure that functions to propel a cell through a liquid medium. [GO:flagellum]

Fold See *Three-dimensional structure of a protein.*

Fossil The mineralized or otherwise preserved remains or traces of an organism. [ENVO: fossil]

Free radical A molecule that is highly reactive because of an unpaired electron, often from an oxygen atom. [CHEBI:radicals]

Freeloader An organism that gains reproductive fitness at the expense of another organism's fitness. Parasites and predators can both be seen as a kind of freeloader.

Function The role that a structure plays in the processes of a living thing; the contribution that a structure makes (perhaps indirectly) to fitness. [GO:biological_process; GO: molecular_function]

Functional group   (of atoms in a molecule) A part of a molecule that tends to produce a characteristic reaction or property; molecules that contain similar functional groups tend to have similar properties.

G protein   A family of guanine-binding proteins that can act as a switch that stays on for a period of time during a cell's processing of a signal. [GO:heterotrimeric G-protein complex]

G-protein coupled receptor   A large family of membrane-bound proteins that detect ligands such as hormones, neurotransmitters, pheromones, odorants, and light and then trigger the activation of G proteins; this family of proteins is the target of many pharmaceutical compounds. [GO:G-protein coupled receptor activity]

G0 phase   The state of a cell that is not actively dividing; such cells might or might not reenter the cell cycle. [GO:chronological cell aging]

G1 phase   The state of a cell after division (M phase) but before DNA synthesis (S phase); the state associated with the majority of cell growth. [GO:G1 phase]

G2 phase   The state of a cell after DNA synthesis (S phase) and before cell division (M phase); this is the shortest phase of cell cycle, dominated by checkpoints that ensure the cell can divide successfully. [GO:G2 phase]

G3P   See *Glyceraldehyde 3-phosphate.*

Gamete   A reproductive cell from a multicellular organism (sperm or egg).

Gap genes   (in development) Transcription factors whose expression defines broad subregions in early embryogenesis. [GO:positive regulation of central gap gene transcription; GO:positive regulation of posterior gap gene transcription; GO:positive regulation of terminal gap gene transcription]

Gap junction   A connection through the membranes of adjacent cells that allows free flow of certain compounds; particularly important in transmitting calcium signals during the contraction of cardiomyocytes. [GO:gap junction]

Gastrula   An early phase of embryogenesis in which cells have formed distinct types and organized into multiple layers.

Gastrulation   The process by which a blastula is transformed into gastrula. [GO:gastrulation]

GDP   See *Guanosine diphosphate.*

Gel electrophoresis   A technique for separating molecules based on their movement through a gel matrix under the influence of an electrical field. [FIX:gel electrophoresis; ECO:inferred from electrophoretic mobility shift assay; SEP:electrophoresis]

Gene   The smallest unit of inheritance; also, a region of DNA that influences a trait. [SO:gene]

Gene duplication   A mutation that results in the addition of an extra copy of at least one gene in a DNA molecule; such events can arise from homologous recombination, chromosomal duplications, and other mechanisms. [SO:intrachromosomal_duplication]

Gene family   A group of genes within an organism that all arose from a single common ancestor and have related functions.

Gene flow   The movement of a gene or genes from one population to another.

Gene product   The protein(s) or RNA(s) that are coded for by a gene. [MI:gene product]

Genetic code   The correspondence between the sequence of a gene and the sequence of the protein specified by that gene.

Genetic discrimination   Treatment or consideration based on genetic status or category rather than individual merit or actual conditions.

Genetic drift   Change in allele frequency from one generation to another within a population resulting from sampling; drift is more extreme in smaller populations.

Genetic locus   (plural genetic loci) A specific region of a DNA sequence that influences a trait; a more precise term than "gene" in a molecular context. [SO:QTL]

Genetic pollution   The unintended transfer of a gene or genes from a genetically engineered organism to other organisms.

Genetic recombination   The formation of new combinations of genes; for example, by crossover during sexual reproduction. [GO:DNA recombination]

Genome   All of the physical hereditary material contained in an organism. [SO:genome]

Genotype   All of the genetic material related to a trait or organism, including those that do not influence the phenotype. [SO:genotype]

Genotyping array   An instrument that determines the genotype of a large set of traits.

Genus   A taxonomic class including more than one species. In the Latin binomial nomenclature used worldwide, the name of an organism is composed of two parts: a genus name (always capitalized) and a species name.

Germ cell   A member of a cell lineage that leads to the production of gametes. [CL:germ cell]

Germ layer   The first specializations of cells during development of a multicellular organism; some organisms form two layers, others three. [GO:formation of primary germ layer]

Germline   In multicellular organisms, cell lineage that has the potential to produce offspring. Contrast with somatic cells. [CL:germ line cell]

GFP   See *Green fluorescent protein.*

Gibbs free energy   The energy in a thermodynamic system that is available to do useful work. [FIX:Gibbs free energy change]

Gland   An organ whose function is to produce a specific substance for excretion or for use elsewhere in the body. [FMA:Gland]

Glia   Specialized cells that surround neurons, providing mechanical support and electrical insulation. [GO:glial cell]

Globular   (protein) A protein that is roughly spherical in shape. Contrast with transmembrane proteins. [PATO:globular]

Glucagon   A pancreatic hormone that increases the concentration of blood sugar by promoting the breakdown of glycogen into glucose; contrast with insulin. [FMA:Glucagon]

Gluconeogenesis   The metabolic processes that produce glucose from noncarbohydrate molecules, such as amino acids or lactic acid. [GO:gluconeogenesis]

Glucose   A six-carbon sugar that is a common intermediate in energy metabolism. [CHEBI:glucose]

Glyceraldehyde 3-phosphate   A three-carbon compound found as an intermediate in several important metabolic pathways. [CHEBI:glyceraldehyde 3-(dihydrogen phosphate)]

Glycine   The chemically simplest amino acid, which also serves as a neurotransmitter; see table 5.1 for its properties. [CHEBI:glycine; CHEBI:glycine residue]

Glycolysis   A nearly universal metabolic pathway in which glucose is converted to pyruvate, producing a net gain of two molecules of ATP. [GO:glycolysis]

Glycoprotein   A protein that has been modified by the covalent attachment of a carbohydrate.

Golgi apparatus   A network of stacked, flattened membranous sacs within the cytoplasm of cells whose major function is to concentrate and package proteins for secretion from the cell. [GO:Golgi apparatus]

GPCR   See *G-protein coupled receptor.*

Green fluorescent protein   A jellyfish protein that spontaneously fluoresces, often fused to other proteins experimentally and used to visualize location and concentration. [ECO:inferred from localization of GFP fusion protein]

Gross anatomy   The study of the structure of the body and its parts without the use of a microscope.

Group selection   A process by which an allele spreads in a population because of the benefits it bestows on a collection of individuals, despite a neutral or negative effect on the fitness of separate individuals.

Growth factor   A protein that functions to stimulate cellular proliferation or differentiation. [GO: growth factor activity]

GTP   See *Guanosine diphosphate.*

Guanine   A nucleotide base found in nucleic acids; pairs with cytosine.[CHEBI:guanine]

Guanosine diphosphate   A doubly phosphorylated nucleoside that plays a role in the G protein signaling pathway. [CHEBI:GDP]

Guanosine triphosphate   A triply phosphorylated nucleoside that plays a role in the G protein signaling pathway. [CHEBI:GTP]

Hairpin turn   (in a protein structure) A portion of a protein structure where the backbone chain sharply changes direction.

Haploid   Having one copy of each chromosome. Contrast with diploid or polyploid. [PATO: haploid]

Hardy-Weinberg equilibrium   A state of a population in which the frequencies of alleles remain the same from generation to generation.

Helicase   A DNA binding protein that functions to unwind the double helix. [GO:helicase activity; GO:DNA helicase complex]

Helper T-cell   A subtype of T-lymphocyte that is required to complete the activation of B-lymphocytes that have recognized an antigen; also functions to transfer information about that antigen to other T-lymphocytes. [CL:T-helper cell]

Hematopoietic stem cell   A stem cell that is the precursor to all types of blood cells. [CL: hematopoietic stem cell]

Heritable   Capable of being transmitted from parent to child.

Heterochromatin   A condensed, transcriptionally inactive form of DNA. [GO:heterochromatin]

Heterotroph   An organism that requires molecules produced by other living things in order to survive. [GO:carbon utilization by utilization of organic compounds]

Heterozygous   Possessing two different alleles for a particular gene. Contrast with homozygous.

Hexokinase   An enzyme that catalyzes the phosphorylation of six-carbon sugars. [GO:hexokinase activity]

High-throughput screening   A process by which a large number of chemical compounds are tested for an activity; widely used in the development of pharmaceuticals.

Histidine   An amino acid; see table 5.1 for its properties. [CHEBI:histidine; CHEBI:histidine residue]

Histology   The study of the microscopic structure of tissues.

Histone   A family of DNA-binding proteins that are the major protein constituent of chromatin in eukaryotes. [CHEBI:histone]

Homeobox genes   A family of genes that together define the anterior-posterior axis in the development of the Bilateria. Also called Hox genes.

Homeostasis   The maintenance of a stable and supportive environment within the body of a multicellular organism. [GO:homeostatic process]

Homogeneous   (mixture) Uniform in composition and properties throughout.

Homologous recombination   A process by which one DNA segment can replace another DNA segment that has a similar sequence. [GO:double-strand break repair via homologous recombination; GO:DNA synthesis during double-strand break repair via homologous recombination]

Homolog   A structure that is similar to another structure because of shared ancestry; often used in reference to genes that descended from a common ancestor. [SO:homologous; SO: homologous_region]

Homozygous   Possessing two identical alleles for a particular gene. Contrast with heterozygous.

Horizontal transmission   (of genes) The exchange of genes between organisms that are not parent and offspring; often through the action of a virus.

Hormone   A chemical substance produced in one part of a multicellular body that functions to convey a signal to another part; usually the product of a gland. [CHEBI:hormone]

Hox genes   See *Homeobox genes*.

Humoral immune response   Antibody production by B-lymphocytes in response to detection of antigens in body fluids.

Hybridize   Form a double-stranded nucleic acid through the hydrogen bonding of complementary nucleotides. [ECO:inferred from nucleic acid hybridization]

Hydrocarbon   An organic molecule consisting of entirely hydrogen and carbon atoms. [CHEBI: hydrocarbons]

Hydrogen bond   A weak bond between a hydrogen atom bonded to an electronegative atom and another electronegative atom; important in the base pairing of nucleic acids and in determining protein structure. [FIX:hydrogen bond]

Hydrolase   An enzyme that catalyzes a reaction involving water, often splitting a large molecule into pieces. [GO:hydrolase activity]

Hydrophobic   Having an aversion to or being repelled by water.

Hydroxide   The anion OH^-, or a compound containing it as a functional group. [CHEBI: hydroxide]

Hyperplasia   An abnormal or unusual increase in number of cells. [PATO:hyperplastic]

Hypertrophy   An abnormal enlargement or growth of an organ or tissue. [PATO:hypertrophic]

Hypothesis-driven   A research methodology in which experiments are designed to test the truth of a specific claim.

Hypoxia   A condition in which there is inadequate oxygen supply to a tissue or organism.

Ichnofossil   A fossil that was never part of an organism, such as a footprint or burrow.

Immune memory   The ability of the adaptive immune system to detect and respond to foreign material more quickly or effectively subsequent to an initial exposure.

Immune system   A collection of cells and organs that recognizes and resists invasion by foreign (nonself) substances and organisms as well as by some cancer cells. [FMA:Immune system]

Immunization   A method to produce immune memory for disease-related antigens without causing disease, usually by inoculation.

Immunoglobulin   See *Antibody*.

Immunoprecipitation   A technique for removing a particular substance from a complex aqueous mixture by using an antibody specific to that substance. [ECO:inferred from immunoprecipitation]

In vitro   Literally, "in glass," meaning outside of an organism (such as in a test tube). [OBI:in-vitro_state; ECO:inferred from in vitro assay]

In vitro fertilization   An assisted reproduction technique in which the combination of egg and sperm occurs outside of the body.

In vivo   In a living organism. [OBI:in-vivo_state; ECO:Inferred from in vivo assay]

Independent assortment   (of genes) The random segregation of alleles to gametes during meiosis. Contrast with linkage.

Inducible knockout   An experimental manipulation that allows the inactivation of a particular gene at a controlled time or location.

Infarction   The death of a portion of tissue from lack of blood supply.

Ingression   The migration of cells from surface layers to the interior of a tissue.

Inhibitory neurotransmitter   A substance released by one nerve cell and detected by another that acts to suppress an action potential or make it less likely that one will be elicited.

Initiation factor   (for ribosomal protein synthesis) One of several proteins that bind to the small subunit of a ribosome at the start of protein synthesis. [GO:translation initiation factor activity]

Innate immunity   A generalized immune response, one that does not give an elevated response on second exposure. Contrast with adaptive immunity. [GO:innate immune response]

Insertion   (in DNA) A mutation that results in the addition of one or more nucleotides in a DNA molecule. [SO:nucleotide_insertion]

Institutional Review Board   An official group associated with an institution performing research on human subjects that acts to ensure that the research is ethical and that the rights of the participants are protected; also called an IRB.

Insulin   A pancreatic hormone that decreases the concentration of blood sugar by promoting the creation of glycogen from glucose, as well as through other metabolic regulation. Contrast with glucagon.

Integumentary system   The cells and tissues that provide the external covering of an organism, including skin, hair, feathers, sweat, and mucus. [FMA:Integumentary system]

Interferon   A cytokine produced in response to viral infection; useful as a pharmaceutical agent against certain viral infections and perhaps some cancers.

Intergenic region   A region of a DNA sequence that does not contain any genes. [SO: intergenic_region]

Interneuron   A neuron that connects other neurons. Contrast with sensory or motor neurons. [CL:interneuron]

Interphase   The portion of cell cycle when the cell is not visibly dividing; includes G1, S, and G2 phases. [GO:interphase]

Interstitial fluid   The internal liquid environment of a multicellular body.

Intracellular   Within a cell. [GO:intracellular]

Intron   A noncoding subsequence within a eukaryotic gene that is removed from the transcribed mRNA before translation into protein. [SO:intron]

Invagination   The infolding of an outer layer so as to form a pocket or pouch. [GO:anterior midgut invagination; GO:posterior midgut invagination]

Inversion   (of DNA) A mutation that causes a chromosomal segment to be reversed. [SO: inversion]

Involution   The inward growth of a group of cells, particularly forming an internal lining of a surface.

Ion   An atom or molecule that has gained or lost electrons, thereby becoming charged. [CHEBI: ions]

Ionic bond   A chemical bond in which electrons are transferred from one atom to another, causing an attraction between the oppositely charged ions. [FIX:ionic bond]

Ischemia   Reduced blood flow to a tissue. [DOID:Ischemia]

Isomers   Compounds that have the same chemical composition, but different structures.

Isomerase   An enzyme that catalyzes a reaction that rearranges the atoms within a molecule, converting one isomer into another. [GO:isomerase activity]

Isotope   One of two or more forms of an element that differ in the number of neutrons in its nucleus; isotopes have the same chemical properties, but different physical properties (particularly mass).

Karyotype   A photomicrograph of metaphase chromosomes, ordered according to size; used for prenatal diagnosis of chromosomal abnormalities and as an aid in identifying species.

Keratin   A tough, fibrous protein that is a major constituent of hair, nails, and hooves. [FMA: Keratin]

Kin selection   A process by which an allele spreads in a population due to the benefits it bestows on relatives, despite a neutral or negative effect on the fitness of an individual.

Kinase   An enzyme that catalyzes the transfer of a phosphate group from a donor, such as ADP or ATP, to an acceptor, usually a protein. [GO:kinase activity]

Kinetics   Properties of a chemical reaction related to the rate at which the reaction takes place. [FIX:kinetic reaction property]

Knockdown   An experimental manipulation that reduces the activity of a particular gene.

Knockout   An experimental manipulation that inactivates a particular gene.

Krebbs cycle   See *Citric acid cycle*. [GO:tricarboxylic acid cycle]

Labyrinth   (middle ear) A complex system of interconnecting cavities in the ear that plays a role in hearing and balance. [FMA:Bony labyrinth; FMA:Membranous labyrinth; FMA:Ethmoidal labyrinth]

Lagging strand   (of DNA in replication) The strand of DNA that is replicated in the 3′ to 5′ direction, and therefore must be synthesized discontinuously during replication. Contrast with the leading strand. [GO:lagging strand elongation]

Laminin   The family of proteins that form the main constituent of basement membranes. [GO:laminin complex; FMA:Laminin]

Landrace   A cultivar of a crop plant that has not been influenced by modern breeding practices; a traditional or preagricultural relative of a crop plant.

Lateral inhibition   A process by which activity in one cell suppresses activity of neighboring cells. [GO:lateral inhibition]

Lead compound   A chemical structure that has properties that suggest it is a good starting point for drug development. [CHEBI:lead molecular entities; CHEBI:organolead compounds]

Leading strand   (of DNA in replication) The strand of DNA that is replicated in the 5′ to 3′ direction, and therefore can be synthesized continuously during replication. Contrast with the lagging strand. [GO:leading strand elongation]

Learning   The ability of an organism to change how it relates sensations to actions through experience of its environment. [GO:learning]

Lesion   An area of diseased or disordered tissue.

Leucine   An amino acid; see table 5.1 for its properties. [CHEBI:leucine; CHEBI:leucine residue]

Leukocyte   See *White blood cell*. [CL:leukocyte]

Ligand   A molecule that binds specifically to another; typically a small molecule that binds to a receptor. [SBO:ligand]

Ligase   An enzyme that catalyzes a reaction that binds two large molecules (such as nucleic acids). [GO:ligase activity]

Limbic system   A set of connected midbrain tissues associated with emotion and memory. [FMA:Limbic system]

Lineage   (in cells) All of the cells that descended asexually from a specific progenitor. [CL:cell by lineage]

Linear polymer   A compound created by the sequential linking of many small chemical units, without branching or cross-linking. [CHEBI:macromolecules]

Linkage   (of genes) A pair of alleles of different genes that are inherited together more frequently than would be expected by chance, as a result of proximity of the genes on a chromosome. Contrast with independent assortment. [SO:linkage_group]

Lipid   A large class of organic molecules that are insoluble in water but soluble in organic solvents. [CHEBI:lipids]

Lipidomics   A large-scale survey of the types and concentrations of lipids in a cell or tissue.

Lipoprotein   A complex molecular assembly in which one or more proteins surround a lipid core; usually functions to transport lipids through aqueous environments such as the blood. [CHEBI:lipoproteins]

Long-term potentiation   An increase in the strength of a synapse that lasts at least several minutes; the basis of some types of learning.

Loop   (of a protein) A relatively flexible portion of a protein backbone external to the core of the protein.

Lyase   An enzyme that catalyzes a reaction involved in the formation or modification of a double bond. [GO:lyase activity]

Lymphocyte   A white blood cell; part of the adaptive immune system. [CL:lymphocyte]

Lysis   (of a cell) The destruction of a cell through damage to its outer membrane. [GO:cytolysis]

Lysosome   A membrane-bound organelle that contains acidic digestive enzymes; functions to break down unnecessary cellular components and foreign materials. [GO:lysosome]

M phase   The part of the cell cycle where the cell divides. [GO:M phase]

Macroevolution   The study of evolutionary changes that take place over a very large timescale, such as speciation or extinction. Contrast with microevolution.

Macromolecules   Large molecules; usually refers to proteins and nucleic acids. [CHEBI: biomacromolecules]

Macrophage   A large white blood cell that engulfs and consumes foreign materials; part of the innate immune system. [CL:macrophage]

Magnetic resonance spectroscopy   A technique used to study the physical, chemical, and biological properties of matter by manipulating its nuclear magnetic resonance. [FIX:magnetic resonance spectroscopy]

Major histocompatibility complex   A large family of proteins responsible for the presentation of antigens on the surface of lymphocytes, and that also play a role in self versus nonself determination. [FMA:Major histocompatibility complex; GO:peptide antigen-transporting ATPase activity]

Malignant   Dangerous to health; typically used to describe a cancerous tumor. [DOID:Malignant Neoplasms]

Mass extinction   A period of time with an especially high rate of species loss.

Mass spectrometry   A method that can rapidly and accurately measure the mass to charge ratio of a large number of ions; used in proteomics, lipidomics, and metabolomics. [FIX:mass spectrometry]

Meiosis   The process of cell division that produces haploid gametes. [GO:meiosis]

Mesenchyme   The embryonic tissue from which all adult connective tissues arises. [CL: mesenchymal cell]

Mesoderm   The middle of the three primary germ layers found in Bilateria. [FMA:Mesoderm; GO:mesoderm development]

Mesosome   An invagination of bacterial cell membranes that anchors a chromosome during cell division. [GO:mesosome]

Messenger RNA   The RNA transcription of a DNA genetic locus; acts as the template that directs a ribosome to synthesize the corresponding polypeptide sequence. Also called mRNA. [SO:mRNA; CHEBI:messenger RNA]

Metabolic pathway   A series of linked, enzymatically catalyzed reactions that transform a substrate into a different, more biologically useful product. [GO:metabolic process]

Metabolism   The chemical breakdown and synthesis of molecules by an organism. [GO: metabolic process]

Metabolite   A substrate in a metabolic pathway.

Metabolomics   A large-scale survey of the types and concentrations of metabolites in a cell or tissue.

Metaphase   The second phase of cell division, when recombined chromosomes are aligned before segregation to opposite poles. [GO:metaphase]

Metaplasia   The process of change of one mature differentiated cell type into another type, abnormal for its context.

Metastasis   The spread of cancer cells from their primary site to other locations in the body. [DOID:Neoplasm Metastasis]

Metazoan   Multicellular organisms whose cells become differentiated to form tissues; all animals except protozoa and sponges are metazoans. [TAXON:Metazoa]

Methylation   (of DNA) The addition of a methyl group to one or more nucleotides; often associated with the inactivation of the genes in a region of DNA. [GO:DNA methylation]

MHC   See *Major histocompatibility complex.*

Micro-RNA   A small, noncoding, single-stranded RNA molecule that plays a role in regulation of gene expression. [SO:miRNA]

Microarray   A two-dimensional grid of many different specific probes (often nucleic acids) in known locations, used to assay the presence or concentration of many molecules simultaneously. [ECO:inferred from array experiment]

Microevolution   The study of evolutionary changes that take place over a relatively short timescale, such as changes of allele frequencies in a population. Contrast with macroevolution.

miRNA   See *Micro-RNA.*

Mitochondrion   (plural Mitochondria) A membrane-bound organelle that is the site of most ATP production in eukaryotic cells; also carries its own genetic material. [GO:mitochondrion]

Mitosis   Asexual cell division. [GO:mitosis]

Mixture   A substance that can be separated into distinct components by physical means; equivalently, a combination of two or more compounds.

Model organism   An organism that can be easily studied in a laboratory to produce results that can be extrapolated to people or other organisms of interest. [ECO:inferred from animal model system]

Modulation   A process that adjusts or changes the activity of another process.

Molarity   A measure of concentration; the number of moles of a solute dissolved in a liter of solution. [UO:unit of molarity]

Mole   A measure of the amount of a substance; the molecular mass of a substance expressed in grams. [UO:mole]

Molecular homology   The property of genes in different organisms of having arisen from a progenitor gene in a common ancestor. [SO:homologous_region; SO:homologous; SO:homologous_to]

Molecular machine   A group of molecules that work together to accomplish a complex biological function. [GO:protein complex]

Molecular mass   The sum of the atomic masses of all of the atoms in a molecule. Also called molecular weight or MW. [FIX:molecular mass]

Molecule   The smallest amount of a compound that retains all of its chemical properties; equivalently, a particular combination of elements bonded together. [CHEBI:molecules]

Monocyte   A large, circulating white blood cell that engulfs and consumes foreign materials; part of the innate immune system; when a monocyte leaves the blood and enters a tissue, it becomes a macrophage. [CL:monocyte]

Monogenic   (trait) A phenotypic quality influenced by only one gene.

Monomer   The smallest repeating unit in a polymer.

Monounsaturated fatty acid   A fatty acid with one double bond. [CHEBI:monounsaturated fatty acids]

Morphogen   A protein that influences a pattern of tissue development, usually by means of a concentration gradient. [GO:morphogen activity]

Morphogenetic movement   Changes in the location of cells during development. [GO:anatomical structure morphogenesis]

Morphogenic field   The distribution of concentrations of morphogens in a developing embryo that defines biological position.

Morphology   The form or structure of an organism or part of an organism. [PATO:morphology]

Morula   An early stage in the development of multicellular organism; a solid ball of cells. [EMAP:TS3, compacted morula; EMAP:TS4, compacted morula]

Motor neuron   A neuron that sends signals to a muscle or gland; also called an efferent neuron. Contrast with interneuron and sensory neuron. [CL:motor neuron]

mRNA   See *Messenger RNA.*

MRS   See *Magnetic resonance spectroscopy.*

Multicellular   An organism with a body consisting of more than a single cell.

Musculoskeletal system   The bones, muscles, ligaments, and tendons; functions to create movement, give shape to the body, and protect internal organs. [FMA:Musculoskeletal system]

Mutant phenotype   The observable characteristics of an organism produced as a result of a mutation. Contrast to wild type. [ECO:inferred from mutant phenotype]

Mutation   A random error in inheritance; also, a change to genomic DNA. [SO:mutation]

Mutualism   A relationship between two kinds of organism living in close association that benefits both.

Mutualist   An organism that lives in close association with another kind of organism, benefiting both. [GO:symbiosis, encompassing mutualism through parasitism]

Myelin   A lipoprotein material that forms a sheathlike covering around nerve fibers. [GO:myelin sheath; FMA:myelin]

Myocardium   The cardiac muscle layer of the heart. [FMA:Myocardium of region of heart]

Myofibril   A bundle of contractile fibers within muscle cells. [GO:myofibril]

Myosin   A family of thick myofibril proteins that, together with actin, causes muscle contraction. [FMA:Myosin heavy chain; FMA:Myosin light chain; GO:myosin complex; GO:myosin filament]

N terminus   (of a polypeptide) The end of a polypeptide or protein that is terminated by a free amine group; also called the amino terminus; conventionally the beginning of the protein, since translation proceeds from N terminus to C terminus. [CHEBI:N-terminal amino-acid residues]

NAD⁺   See *Nicotinamide adenine dinucleotide.*

NADP⁺   See *Nicotinamide adenine dinucleotide phosphate.*

NADH   The reduced form of NAD⁺. [CHEBI:NADH]

NADPH   The reduced form of NADP⁺. [CHEBI:NADPH]

Natural killer cell   A cytotoxic lymphocyte that forms a major part of the innate immune system. [CL:natural killer cell]

Natural selection   Differential reproductive success due to heritable factors; one of the three central aspects of evolution.

ncRNA   See *Noncoding RNA.*

Necrosis   A form of cell death that results in the rupture of cell membranes and the leakage of cytoplasm into the interstitial fluid. [GO:cell death]

Negative feedback loop   A process by which deviations from a norm are detected and corrected, tending to keep a system in a steady state. [GO:negative regulation of biological process]

Nervous system   The totality of nerves and glia in the body, including the brain, spinal cord, and sensory and motor neurons; functions to sense the external world, evaluate situations, coordinate motion, and control many homeostatic parameters. [FMA:Nervous system]

Neural crest   A region of neural precursor cells in an embryo that remain outside of the developing neural tube; these cells migrate widely throughout the embryo. [FMA:Neural crest]

Neuromuscular junction   The connection between a motor nerve and the muscle it innervates. [GO:neuromuscular junction]

Neuron   An excitable cell that functions to receive, process, and transmit information. [CL: neuron]

Neurotransmitter   A large class of molecules that function to transmit information from one neuron to another across a synapse. [CHEBI:neurotransmitter]

Neutral mutation   A heritable change that has no effect on fitness. [SO:silent_mutation]

Neutron   An uncharged subatomic particle, typically found in the nucleus of an atom. [CHEBI:neutron]

Neutrophil   The most common type of white blood cell; part of the innate immune system. [CL:neutrophil]

Niche   The position or role occupied by a particular species in an ecosystem, both the physical spaces that it inhabits and the functions that it performs within a community.

Nicotinamide adenine dinucleotide   An important coenzyme derived from adenine and the B vitamin niacin that functions as a hydrogen carrier in a wide range of redox reactions; usually written NAD⁺. [CHEBI:NAD(+)]

Nicotinamide adenine dinucleotide phosphate   A coenzyme similar to NAD+ and present in most living cells but serves as a hydrogen carrier in different metabolic processes; usually written NADP⁺. [CHEBI:NADP(+)]

NMR   See *Nuclear magnetic resonance.*

Noncoding RNA   RNA molecules that do not act as messenger RNA, such as ribosomal or small interfering RNAs. Also called ncRNA. [SO:ncRNA]

Nonpolar   (molecule) A molecule with an even distribution of charge over its surface; nonpolar molecules are generally hydrophobic. [FIX:non-polar covalent bond]

Nuclear magnetic resonance   A quantum mechanical phenomenon based on the fact that atomic nuclei spin around the axis of a strong magnetic field; spinning nuclei create oscillating magnetic fields and emit a detectable amount of electromagnetic radiation. [FIX:nuclear magnetic resonance spectroscopy]

Nucleic acid   Polymers of nucleotides; DNA and RNA. [CHEBI:nucleic acids]

Nucleolus   An organelle found inside the nucleus of many eukaryotic cells; functions as the location of transcription of rRNA molecules and the assembly of ribosomal subunits. [GO:nucleolus]

Nucleoside   A dinitrogen heterocycle bonded to a five-carbon sugar; a nucleotide without the phosphate group. [CHEBI:nucleosides]

Nucleotide   A dinitrogen heterocycle bonded to a five-carbon sugar and a phosphate group; the monomeric unit of the nucleic acids. [CHEBI:nucleotides]

Nucleus   (plural nuclei) A membrane-bound organelle of eukaryotic cells that contains the chromosomes and is the location of transcriptional regulation and mRNA processing; typically, a spherical, central structure that is the largest organelle in a cell. [GO:nucleus]

Okazaki fragments   A relatively short fragment of DNA with an RNA primer at its 5′ terminus, created on the lagging strand during DNA replication. [GO:DNA replication, Okazaki fragment processing]

Oncogene   A gene that is capable, particularly when damaged or mutated, of causing the transformation of normal cells into cancer cells.

Orbital   A mathematical description of the region within an atom that has the highest probability density for an electron.

Organ   A tissue or group of tissues that constitute a morphologically and functionally distinct part of an organism. [CARO:simple organ; CARO:complex organ; FMA:Organ]

Organ system   A group of organs that have related functional roles. [FMA:Organ system]

Organelle   A subcellular structure, usually bound by a membrane, that performs a specific function. [GO:organelle]

Organogenesis   The time period during embryonic development in which organs and organ systems are formed. [GO:organ development]

Osmotic pressure   Pressure generated by water moving across a cell membrane to equalize solute concentrations on either side of the membrane.

Oxaloacetate   A four-carbon compound that plays a central role in the TCA cycle. [CHEBI: oxaloacetate(2-)]

Oxidation-reduction   (reaction) A reaction that involves transfer of an electron from one substance to another, with a consequent change in oxidation states. [REX:redox reaction]

Oxidative phosphorylation   The formation of ATP from ADP using energy derived through an electron transport chain, typically from NADH to molecular oxygen. [GO:oxidative phosphorylation]

Oxidative stress   Adverse effects occurring when the generation of reactive oxygen species in a system exceeds the system's ability to neutralize or eliminate them; this can result in damage to lipids, proteins, and DNA.

Oxidize   (a compound) To take an electron away, usually by reaction with oxygen, changing the oxidation state of the compound. [REX:oxidation]

Oxidoreductase   An enzyme that catalyzes oxidation-reduction reactions. [GO:oxidoreductase activity]

Pair-rule genes   (in development) Developmental genes whose products are expressed in alternating stripes and that function to specify different segments of the body. [GO:periodic partitioning by pair rule gene]

Pancreas   An organ near the stomach that functions to secrete pancreatic juices into the gastrointestinal tract and insulin and glucagon into the blood. [FMA:Pancreas]

Paralog   Homologous genes within a single organism; paralogs arise by gene duplication, usually followed by divergence in function. [SO:paralogous_region; SO:paralogous; SO: paralogous_to]

Parasite   An organism that lives in close association with another, benefiting (often by feeding) at the other's expense. [GO:symbiosis, encompassing mutualism through parasitism; PLO:parasite]

Particulate   (inheritance) An all-or-none transmission of small units of genetic information from one parent or the other. Contrast with blending inheritance.

Pathogen   An organism that causes disease.

Pathogenesis   The origin and development of a disease. [GO:pathogenesis]

PCR   See *Polymerase chain reaction.*

Peptidase   An enzyme that catalyzes the hydrolysis and consequent breakage of peptide bonds. [GO:peptidase activity]

Peptide bond   A covalent bond between the amino nitrogen in one amino acid and the carboxy carbon in another that links them together into a polypeptide chain. [GO:peptide biosynthetic process]

Periodic table of the elements   An arrangement of elements in a geometric pattern that aligns elements based on the number of orbitals (rows) and the number of electrons filling the outermost orbital (columns).

Peritoneum   The membrane that lines the coelom, and surrounds many of the organs within it. [FMA:Peritoneum]

Peroxisome   A membrane-bound organelle that contains oxidative enzymes and functions in beta oxidation and the detoxification of free radicals. [GO:peroxisome]

Perturbation   A small change to a system in equilibrium; in engineering, often used to study the dynamics of complex systems.

Phage   A virus that infects bacteria.

Phagocyte   A cell of the innate immune system that can surround and ingest foreign matter or debris. [CL:phagocyte]

Pharmacogenomics   The study of the effect of individual human genetic variation on response to medication.

Pharmacophore   The molecular properties that are the essential features responsible for a drug's biological activity. [SOPHARM:pharmacogenomic property]

Phenotype   The entirety of observable characteristics of an organism. [PATO:physical quality]

Phenotypic plasticity   The ability of an organism with a given genotype to change its phenotype in response to changes in the environment.

Pheophytin   A kind of chlorophyll that acts as an electron acceptor in the type II photosystem. [CHEBI:pheophytins]

Pheromone   A substance secreted by an organism to affect the behavior or development of other members of the same species. [CHEBI:pheromone]

Phi angle   (of a peptide bond) The dihedral angle that describes rotation around the alpha carbon to nitrogen bond within an amino acid.

Phosphate   A functional group consisting of an atom of phosphorus bound to four atoms of oxygen ($-PO_4$); called inorganic phosphate when not bound to another compound.

Phosphoglucose isomerase   An enzyme that catalyzes a key step in glycolysis, and is also involved in gluconeogenesis. [GO:glucose-6-phosphate isomerase activity]

Phospholipid   A lipid with at least one phosphate group in the hydrophilic part of the molecule; the major constituent of cell membranes. [CHEBI:phospholipids]

Phosphorylation   The chemical addition of a phosphate group to a compound. [GO:phosphorylation]

Photoreceptor   A light-sensitive neuron. [CL:photoreceptor cell]

Photosynthesis   The process by which green plants, algae, and some bacteria absorb light and use it to produce ATP and synthesize organic compounds from carbon dioxide and water. [GO:photosynthesis]

Photosystem   A complex of proteins and other molecules involved in photosynthesis. [GO:photosystem I; GO:photosystem II]

Physiology   The study of bodily functions.

Plasma   The liquid part of the blood, lymph, and intracellular fluid in which cells are suspended. [FMA:Plasma]

Plasma cell   A fully differentiated B-lymphocyte that has been activated by an antigen and is producing antibodies. [CL:plasma cell]

Plasmid   A genetic particle, usually circular DNA, that is physically separate from the chromosome of the host cell (usually a bacterium) and that can stably function and replicate; it is not essential to the cell's basic functioning. [SO:plasmid]

Plasticity   The ability to change; in neurons, the changes in the strength of synaptic connections resulting from experience. [GO:regulation of synaptic plasticity]

Platelet   A small blood cell that plays a central role in the form of blood clots. [CL:platelet]

Pleiotropy   The quality of a single gene having effects on several seemingly unrelated traits.

Pluripotent   Able to differentiate into many types of cells; often said of embryonic stem cells that can differentiate into all cell types except placental cells.

Polar molecule   A molecule with an uneven distribution of charge over its surface; polar molecules are generally water soluble. [FIX:polar covalent bond]

Poly-A tail   A sequence of repeating adenosine nucleotides at the 3′ end of an mRNA. [SO:polyA_sequence]

Polyubiquitination   The binding of many ubiquitin molecules to the same protein, a signal that leads to degradation of the protein in the proteasome. [GO:protein polyubiquitination]

Polygenic   (trait) Pertaining to or influenced by several genes.

Polymerase   An enzyme that catalyzes the creation of polymers, particularly the addition of nucleotides to nucleic acids. [GO:DNA-directed DNA polymerase activity]; [GO:DNA-directed RNA polymerase activity; GO:RNA-directed DNA polymerase activity; GO:RNA-directed RNA polymerase activity]

Polymerase chain reaction A fast, inexpensive laboratory technique for making a large number of copies of a particular DNA sequence. [ECO:inferred from PCR experiment; FIX:nucleic acid sequencing using polymerase]

Polymorphism A quality that exists in more than one form or type; something that differs among organisms. [SO:sequence_alteration; ADW:polymorphism]

Polypeptide A linear polymer of amino acids connected by peptide bonds; a large polypeptide is called a protein. [CHEBI:polypeptides]

Polyprotein Several proteins synthesized as a single polypeptide that must be cleaved into smaller pieces before becoming active.

Polyunsaturated fatty acid A fatty acid with two or more double bonds. [CHEBI:polyunsaturated fatty acids]

Population A group of closely related organisms living in proximity to each other.

Population genetics The study of allele frequency distribution and change in populations.

Pore (in a membrane) A group of proteins that make an opening in a membrane through which liquids or other molecules can flow. [GO:pore complex]

Postsynaptic On the receiving (dendrite) side of a connection between two neurons. [GO: postsynaptic membrane; FMA:Postsynaptic component of chemical synapse; FMA:Axon of postganglionic neuron]

Posttranslational modification (of a protein) Covalent modification of a protein following translation, for example, the addition of a phosphate group. Also called a PTM. [GO:post-translational protein modification]

Posterior See *Anterior-posterior axis.*

Pre-mRNA The immediate result of eukaryotic transcription, before processing such as the addition of a poly-A tail or the removal of introns. [SO:pre_edited_mRNA]

Precautionary principle A moral and political principle stating that if an action or policy might cause severe or irreversible harm to the public, in the absence of a scientific consensus that harm would *not* ensue, the burden of proof falls on those who would advocate taking the action.

Preclinical research Studies in vitro and in model organisms to establish an expectation of safety and efficacy for a potential pharmaceutical before beginning tests in humans.

Primary structure (of a protein) See *Amino acid sequence.*

Primordia An organ or tissue in its earliest recognizable stage of development. [CL:primordial germ cell]

Product (of a reaction) A substance formed as a result of a chemical reaction [SBO:product]

Prokaryote A single-celled organism lacking a nucleus; a bacterium. [TAXON:Bacteria]

Proliferation An increase, generally rapid, in the number of cells in a tissue through mitosis. [GO:cell proliferation]

Proline An amino acid; see table 5.1 for its properties. [CHEBI:proline; CHEBI:proline residue]

Promoter A region of DNA to which RNA polymerase can bind, initiating transcription; the region just before (toward the 5′ end of) a coding sequence. [SO:promoter]

Prophase The first stage of cell division (both mitotic and meiotic), when the duplicated chromosomes condense, are physically separated from each other, and recombination occurs. [GO:prophase]

Proprioceptor A sensory nerve ending that responds to changes in tension in a muscle or tendon. [GO:proprioception]

Protein A linear polymer of amino acids; the class of molecule coded for by most genes; the molecule that catalyzes most biological functioning. [CHEBI:proteins; CHEBI:protein polypeptide chains]

Protein folding problem See *Protein structure prediction.*

Protein sequence See *Amino acid sequence.*

Protein structure prediction The task of estimating or calculating the three-dimensional structure of a protein from its amino acid sequence alone.

Proteomics A large-scale survey of the types and concentrations of proteins in a cell or tissue.

Proteasome A large multiprotein complex that is a site of degradation for ubiquitinated proteins. [GO:proteasome complex (sensu Eukaryota)]

Protist A large class of single-celled eukaryotes.

Proton A particle with positive charge in the nucleus of atoms. [CHEBI:proton]

Proximal-distal axis Imaginary lines running from the center of an organism to its peripheries; the dimension of closer versus further from the center. [GO:proximal/distal axis specification; PATO:proximal to; PATO:distal to]

Pseudogene A nonfunctional DNA sequence descended from a functional ancestor, often arising by gene duplication and function loss through mutation. [SO:pseudogene]

Psi angle (of a peptide bond) The dihedral angle that describes rotation around the alpha carbon to carbonyl carbon bond within an amino acid.

Public health An approach to medicine that is concerned with the health of a community as a whole. Contrast to medical health care, which focuses on treatment of individuals.

Pulmonary circulation The portion of the cardiovascular system that carries oxygen-depleted blood away from the heart, to the lungs, and also returns oxygenated blood back to the heart. Contrast with systemic circulation. [FMA:Pulmonary vascular system]

Punctuated equilibrium An evolutionary theory that postulates new species appear suddenly (within a few hundred thousand years) in the fossil record and then show little change for millions of years until their extinction.

Pure (substance) Unable to be separated into distinct constituents by chemical means. [SEP: substance]

Purifying selection See *Directional selection.*

Pyruvate A three-carbon compound that is the principle product of glycolysis and a participant in many other metabolic pathways. [CHEBI:pyruvate]

Quaternary structure (of a protein) The assembled structure of two or more polypeptide chains that function as that assembled unit. [FIX:protein quaternary structure]

Reactants The substances that interact in a chemical reaction. [SBO:reactant]

Reaction (chemical) The chemical interaction of two or more substances to form one or more new substances. [REX:chemical reaction; SBO:reaction]

Reaction rate The change in concentration over time of reactants and products in a chemical reaction. [SBO:kinetic constant]

Reactive oxygen species Free radicals that contain oxygen.

Receptor A molecule, usually a membrane-bound protein, that can be activated by binding to a specific ligand. [GO:receptor activity]

Receptor tyrosine kinase A family of transmembrane protein receptors that when activated catalyzes the phosphorylation of a tyrosine, often within another receptor tyrosine kinase. [GO: receptor signaling protein tyrosine kinase activity]

Recessive (allele) An allele that influences phenotype only in homozygous organisms. Contrast to dominant.

Recognition site (of an enzyme) The portion of an enzyme responsible for the specificity of interaction with its ligand. [SO:binding_site]

Recombinant DNA Biologically active DNA created by the artificial combination of segments of DNA from different sources, often from different species. [SO:recombination_feature]

Recombination The natural process in which homologous chromosomes exchange segments, usually during meiosis, creating offspring that are a random combination of parental alleles. [GO: somatic recombination of immunoglobulin gene segments; GO:somatic recombination of T-cell receptor gene segments]

Red blood cell A cell that arises in the bone marrow, circulates in the blood, and expresses large amounts of hemoglobin, functioning to carry oxygen. Also called an erythrocyte. [CL: erythrocyte]

Redox See *Oxidation-reduction reaction.* [REX:redox reaction]

Reduce (a compound) To add an electron, changing the oxidation state of the compound. [REX: reduction]

Reflex A simple, direct connection between a sensation and an action, often requiring only a single interneuron.

Regulation A process that controls the activity of a molecule or of another process.

Regulatory sequence See *Cis regulatory sequence.*

Replicate Create a copy, usually in reference to nucleic acid sequences.

Replication fork The point at which the two strands of a DNA molecule are separated during DNA duplication. [GO:replication fork]

Repressor (of gene expression) A regulatory sequence that tends to decrease the expression of an associated coding sequence. [GO:transcriptional repressor complex]

Reproduce Create an offspring, another organism that has traits similar to those of its parent(s). [GO:reproduction]

Reproductive isolation A state of a population, often resulting from a geographic barrier, that prevents the exchange of genes with any other population.

Reproductive success The number of fertile offspring an organism produces, often defined relative to other organisms with varying genotypes. [PATO:reproductive quality]

Reproductive system The organs and tissues involved in the production and maturation of gametes, in their union and in their subsequent development as offspring. [FMA:Genital system]

Residue (of an amino acid) The portion of an amino acid remaining after peptide bonding; the constituents of a polypeptide. [CHEBI:amino-acid residues]

Respiratory system The organs and tissues that function to regulate gases, primarily oxygen and carbon dioxide, dissolved in the interstitial fluid. [FMA:Respiratory system]

Restriction enzyme A protein that recognizes specific nucleotide sequences in a DNA molecule and cuts the molecule at that points; used by bacteria to combat infection by viruses, and by genetic engineers to manipulate DNA. [GO:restriction endodeoxyribonuclease activity]

Restriction-modification system A set of proteins in bacteria that combat infection by viruses, including restriction enzymes and other enzymes that protect potential target sequences in the bacterium's own DNA by methylation. [GO:DNA restriction-modification system]

Resuscitation Medical procedures undertaken to ensure that the respiratory and cardiovascular systems are able to function well enough to prevent immediate death.

Retina The photoreceptor-containing tissue in an eye. [FMA:Retina]

Reversible (reaction) A chemical reaction in which products and reactants can exchange roles.

Ribonucleic acids Linear polymers of ribonucleotides whose biological functions include transmitting information, catalysis, and regulation of gene expression; also called RNA.

Ribonucleotide A dinitrogen heterocycle bonded to a ribose and a phosphate group; the monomeric unit of ribonucleic acids.

Ribose A five-carbon sugar molecule. [CHEBI:ribose; CHEBI:riboses]

Ribosomal RNA The RNA molecules that are essential structural and functional components of ribosomes; also called rRNA. [SO:rRNA; CHEBI:ribosomal RNA]

Ribosome A cytoplasmic organelle composed of protein and RNA in which protein synthesis occurs. [GO:ribosome]

Ribozyme An RNA molecule that functions as a catalyst. [SO:ribozyme]

Ribulose bisphosphate   An important 5-carbon intermediate in the Calvin cycle. [CHEBI:D-ribulose 1,4-bisphosphate]

RNA   See *Ribonucleic acids.*

RNA interference   A mechanism of RNA-based regulation of gene expression in which a double-stranded RNA triggers the destruction of single-stranded RNA molecules that contain the same sequence. [GO:RNA interference]

RNA polymerase II   A complex of proteins that catalyzes the transcription DNA resulting in the synthesis of pre-mRNA and most snRNA. Also called Pol II. [GO:DNA-directed RNA polymerase II, core complex; GO:DNA-directed RNA polymerase II, holoenzyme]

RNA polymerase III   A complex of proteins that catalyzes the transcription DNA resulting in the synthesis of most transfer, ribosomal, and other small RNAs. [GO:transcription from RNA polymerase III promoter]

RNA-induced silencing complex   A complex of enzymes that cleaves RNA molecules that match a double-stranded RNA template; one of the mechanisms of RNA interference. [GO:RNA-induced silencing complex]

RNAi   See *RNA interference.*

rRNA   See *Ribosomal RNA.*

RTK   See *Receptor tyrosine kinase.*

RuBisCO   A common short name for ribulose-1,5-bisphosphate carboxylase/oxygenase, an enzyme that plays a central role in the Calvin cycle. [GO:ribulose-bisphosphate carboxylase activity]

S phase   The period during the cell cycle when DNA is synthesized. [GO:S phase]

Saturated hydrocarbon   A hydrocarbon in which all of the carbons are bonded by single bonds.

Secondary messenger   A molecule within a cell that is responsible for initiating or maintaining the response to a signal from another chemical messenger (such as a hormone) that does not enter the target cell itself. [GO:second-messenger-mediated signaling]

Secondary metabolism   The metabolic activities seen only in a specialized subset of organisms. Contrast with core metabolism. [GO:secondary metabolic process]

Secondary structure   (of a protein) The composition of a protein described in terms of common local substructures, such as alpha helices or beta sheets. [SO:polypeptide_secondary_structure; FIX:protein secondary structure]

Segregation   (of chromosomes) The physical separation of paired chromosomes (and hence of alleles) from each other during meiosis. [GO:chromosome segregation]

Selectively permeable   (membrane) A quality of a membrane that allows some compounds, but not others, to pass through.

Selex   A biotechnology that uses in vitro selection to produce ribozymes with a desired function.

Selfish genes   Alleles that can propagate themselves at the expense of (some of) the organisms that carry them.

Senescence   A state in which growth ceases; in cells, a state in which the cell no longer divides. [GO:senescence]

Sensory neuron   A nerve cell whose activity is responsive to stimuli external to nervous system. Contrast to interneuron and motor neuron. [CL:sensory neuron]

Sequence   (of a polymer) An ordered list of the monomers that constitute a linear polymer. [FIX: primary sequence]

Serial homologs   Repeated structures, derived from a single ancestor; an important constituent of many animal bodies.

Serine   An amino acid; see table 5.1 for its properties. [CHEBI:serine; CHEBI:serine residue]

Serine protease   An enzyme that catalyzes the cleavage of another protein through a serine residue in its active site. [GO:serine-type endopeptidase activity]

Sexual selection   A type of natural selection that acts differently on males and females of the same species; often seen in traits that influence mating choices or attractiveness to potential mates.

SH2 domain   A region found in many proteins that typically recognizes and binds to a specific phosphorylated tyrosine residue; often part of adaptor proteins in signal transduction pathways. [GO:SH3/SH2 adaptor activity]

SH3 domain   A region found in many proteins that typically mediates assembly of specific protein complexes by binding to specific proline-rich peptides; often part of adaptor proteins in signal transduction pathways. [GO:SH3/SH2 adaptor activity]

Side chain   (of an amino acid) The portion of an amino acid that is not part of the polypeptide backbone and varies among the different amino acids.

Side chain angle   (of an amino acid) The angle that specifies the rotation around the bond that attaches a side chain to the backbone.

Signal sequence   (for protein localization) A short amino acid sequence, usually at the N terminus of a protein, that causes the protein to be transported to a particular organelle or cellular compartment. [SO:signal_peptide]

Signal transduction pathway   A series of chemical transformations by which a signal outside of a cell causes a functional change inside that cell. [GO:signal transduction]

Silent mutation   A kind of neutral mutation in a coding sequence that results in a codon that codes for the same amino acid as the original. [SO:silent_mutation]

Single nucleotide polymorphism   A difference of a single nucleotide among DNA sequences. [SO:SNP]

siRNA   See *Small interfering RNA*.

SMAD   A family of signal transduction molecules that is responsive to ligands from the transforming growth factor beta (TGF-β) superfamily.

Small interfering RNA   Endogenously produced double-stranded RNA molecules of 20 to 25 nucleotides that play a role in regulation of gene expression through RNA interference. [SO:siRNA]

Small molecule   In biology, any compound that is not a protein or nucleic acid. Contrast with macromolecule.

SNP   See *Single nucleotide polymorphism*.

Solute   The substance(s) in a solution present in the lesser amounts; often a solid that has been dissolved in a liquid.

Solution   A homogeneous mixture of two or more components; often one or more solutes dissolved in a liquid solvent. [SEP:solution]

Solvation effect   An electrostatic attraction between the atomic charges holding a solute molecule together and the polar water molecules in an aqueous solution. [REX:solvation]

Solvent   The substance present in a solution in the greatest amount; often a liquid in which solute(s) are dissolved.

Soma   In a multicellular organism, the cells that are not part of the germline. Also called somatic cells. [CL:somatic stem cell]

Southern blot   A method for transferring a DNA molecule that has been purified by gel electrophoresis. [FIX:Southern blotting; ECO:inferred from Southern blot hybridization; SEP: Southern blot]

Species   A group of potentially interbreeding organisms that are reproductively isolated from other such groups; see also taxon.

Spinal cord   The portion of the central nervous system that conducts sensory and motor nerve impulses to and from the brain; a long tubelike structure extending from the base of the brain through the vertebral canal. [FMA:Spinal cord]

Spindle   (in cell division) A fibrous protein network formed during cell division that functions to move duplicated chromosomes to opposite poles within a cell. [GO:spindle]

Spliceosome   A large complex of proteins and RNA molecules in the nucleus of eukaryotic cells that functions to remove introns from pre-mRNA. [GO:spliceosome]

Statolith   A plant organelle that senses gravity.

Stem cell   A cell that functions to produce differentiated cells; the only source of terminally differentiated cells. [CL:stem cell]

Steroids   A large family of structurally similar lipids that contain four interconnected carbon rings. [CHEBI:steroids]

Stoichiometry   The quantitative relationships among compounds in a reaction.

Stop codon   A triplet of nucleotides that specifies the end of a coding sequence and signals the ribosome to release a newly synthesized polypeptide. [SO:stop_codon]

Strand   (of DNA) A single DNA polymer, not bonded to a complementary polymer. [CHEBI: single-stranded DNA]

Structure   The physical components of a living system and how they relate to each other; what a thing is. [CARO:anatomical structure; SO:sequence_secondary_structure; PATO:structure; FIX: structure]

Substrate   The substance on which an enzyme acts. [SBO:substrate]

Superfamily   (of proteins) A group of functionally related proteins that are not necessarily homologous. [MI:scop superfamily]

Synapse   A junction between a neuron and either another neuron or a muscle that transmits chemical signals from one cell to the next. [GO:synapse; FMA:Synapse]

Synthetic lethal   A type of experiment that demonstrates two gene products interact by showing that knockouts in both, but not in either alone, prevent reproduction.

Systemic circulation   The portion of the cardiovascular system that carries oxygenated blood from the heart to the body, and oxygen-depleted blood back to the heart. Contrast with pulmonary circulation. [FMA:Systemic arterial system; FMA:Systemic venous system]

Systems biology   The study of the interactions among molecular components of a biological system, and how these interactions give rise to function.

Systole   The period of time when the heart ventricles contract.

T-cell   The type of white blood cell responsible for cell-mediated immunity, also called a T-lymphocyte [CL:T-cell]

Taxon   (plural taxa) A group or category of living organisms linked by common ancestry; similar to a species but also applies to organisms that reproduce asexually.

TCA cycle   See *Citric acid cycle.*

Telomere   The end of a chromosome, contains repetitious DNA sequences. [GO:telomere cap complex; GO:chromosome, telomeric region]

Telophase   The final phase of eukaryotic cell division, when membranes form around each set of chromosomes, creating two nuclei. [GO:telophase]

Template-directed DNA repair   A mechanism that restores damaged DNA by copying from the complementary, undamaged strand.

Terminal differentiation   The process that results in a fully mature cell type, usually one that does not further divide.

Termination factor   (for ribosomal protein synthesis) A set of enzymes involved in freeing a newly synthesized polypeptide chain from a ribosome, and preparing the ribosome to accept another mRNA. [GO:translation termination factory activity]

Terpenes   A large and varied class of hydrocarbons, produced primarily by plants. [CHEBI:terpenes]

Tertiary structure   (of a protein) See *Three-dimensional structure.*

TGF-β   See *Transforming growth factor beta superfamily.*

Thermodynamically feasible reaction   A chemical reaction that would result in a reduction of Gibbs free energy.

Thermodynamics   The physical theory of different forms of energy and their interconversion. [FIX:thermodynamic reaction property]

Three-dimensional structure   (of a protein) The precise location of every atom in a protein; equivalently, the precise set of phi, psi, and side chain angles of a polypeptide chain. [FIX: conformation; FIX:protein tertiary structure]

Thrombocyte   See *Platelet*.

Thylakoids   Membrane-bound compartments within chloroplasts that are the location that photosynthesis takes place. [GO:thylakoid]

Thymine   A nucleotide base found in nucleic acids; pairs with adenine. [CHEBI:thymine]

Tissue   A group of similar cells, specialized to perform a particular function. [CARO:portion of tissue; FMA:Tissue]

Topoisomerase   A class of isomerases that manipulate the topology, particularly the coiling, of DNA. [GO:DNA topoisomerase activity]

Totipotent   (stem cell) Able to differentiate into any different type of cell. [CL:totipotent stem cell]

Trait   An inherited characteristic of an organism. [PATO:quality]

Trans   Latin for "across from" or "far away." For a gene, it means something (usually a protein) that comes from far away to act on the gene.

Transamination   A chemical reaction that transfers a nitrogen-containing group from one molecule to another.

Transcript   See *Messenger RNA*.

Transcription   The creation of a messenger RNA from a coding sequence. [GO:transcription]

Transcription factor   A DNA-binding protein that influences the transcription of a particular coding sequence. [GO:transcription factor complex]

Transcription factor binding site   A cis regulatory site in DNA that is recognized and bound to by a transcription factor. [SO:TF_binding_site]

Transcriptional control   The set of mechanisms that determines which coding sequences are turned into messenger RNA at any given time, and at what rate. [GO:regulation of transcription]

Transfer RNA   A type of RNA that bonds at one end to a specific amino acid and at another hybridizes to the corresponding codon on an mRNA; the physical realization of the genetic code. [SO:tRNA; CHEBI:transfer RNA]

Transferase   An enzyme that catalyzes the transfer of a functional group from one molecule to another. [GO:transferase activity]

Transforming growth factor beta superfamily   A large and diverse family of intercellular signaling proteins, including many growth factors and morphogens. [GO:transforming growth factor beta receptor activity]

Transgene   Genetic material that has been artificially transferred from one organism to another. [SO:transgene]

Translation   The process by which genetic information coded in messenger RNA directs the formation of a specific protein at a ribosome. [GO:translation]

Transposition   (of DNA) A kind of mutation in which a chromosomal segment is moved to a new position. [GO:transposition]

Transposon   A naturally occurring DNA sequence that is capable of catalyzing its own duplication or the movement of its location within a genome. [SO:transposable_element]

Tricarboxylic acid cycle   See *Citric acid cycle*.

Triglyceride   A compound consisting of three fatty acids covalently bonded to a glycerol; the most common form of fats and oils used for energy storage. [CHEBI:triacylglycerols]

tRNA   See *Transfer RNA*.

Tryptophan   An amino acid; see table 5.1 for its properties. [CHEBI:tryptophan; CHEBI: tryptophan residue]

Tumor   An abnormal mass of undifferentiated cells within a multicellular organism; may be benign or malignant. [DOID:cancer]

Tumor suppressor gene   A gene whose product functions to detect DNA damage or otherwise provide defense against uncontrolled growth.

Turn   (in a protein structure) A large change in the direction of a protein backbone chain. [SO:turn]

Ubiquitin   A small protein that, when covalently attached to other proteins, signals that the other protein is to be degraded. [GO:protein ubiquitination; GO:ubiquitin-protein ligase activity]

Unipotent   A stem cell that can produce only one cell type.

Uracil   A nucleotide base found in RNA; pairs with adenine. [CHEBI:uracil]

Urinary system   A collection of tissues and organs that functions to maintain homeostasis by regulating water balance and removing harmful substances from the blood. [FMA:Urinary system]

UTR   See *5′ untranslated region or 3′ untranslated region.*

Valence   The number of electrons in the outermost shell of an atom; roughly, a measure of the number of chemical bonds that can be formed by atoms of a particular element.

Van der Waals (VdW) force   A weak physical force that acts on molecules and atoms in close proximity to each other; it is repulsive at short distances and attractive at longer ones. [FIX:Van der Waals bond]

Vasoconstriction   Narrowing of the internal diameter of blood vessels due to contraction of the smooth muscles in their walls. Contrast with vasodilation. [GO:vasoconstriction]

Vasodilation   Widening of the internal diameter of blood vessels by relaxation of the smooth muscles in their walls. Contrast with vasoconstriction. [GO:vasodilation]

Vector   (for genetic engineering) A mechanism for adding an engineered DNA sequence to the genome of an organism. [SO:vector]

Vein   A blood vessel that returns blood to the heart. [FMA:Vein]

Ventral   See *Dorsal-ventral axis.*

Ventral nerve cord   A collection of neurons that makes up the central nervous system of many invertebrates. [FBbt:ventral nerve cord]

Ventricle   A cavity within an organ; especially the cavities in the heart that contain blood to be pumped and cavities in the brain that contain cerebrospinal fluid. [FMA:Ventricle]

Vesicle   A small, intracellular, membrane-bounded sac in which substances are transported or stored. [GO:vesicle]

Visual cortex   A brain region that integrates and processes information from the retina. [FMA: Striate cortex]

Voltage-gated channels   Proteins channels through a membrane that open or close depending on the voltage across the membrane. [GO:voltage-gated channel activity]

White blood cell   A cell of the immune system that circulates in the blood; also called a leukocyte. [CL:leukocyte]

Wild type   The typical form of an organism, strain, gene, or characteristic as it occurs in nature.

X-ray crystallography   A method that analyzes the diffraction patterns of X-rays through a crystal of a protein to determine the three-dimensional structure of that protein. [FIX:X-ray crystallography]

Yeast 2-hybrid   An experimental method that can rapidly test a large number of proteins for binding affinity to another protein. [ECO:inferred from yeast 2-hybrid assay]

Z-scheme   The biochemical mechanism by which light is transformed into energy. [GO: photosynthesis, light reaction]

Zygote   The diploid cell resulting from the union of two haploid cells of complementary mating types; a fertilized egg. [CL:zygote]

Index

Figures are indicated by italic; footnotes are indicated by n. followed by note number.